WORK FLOW

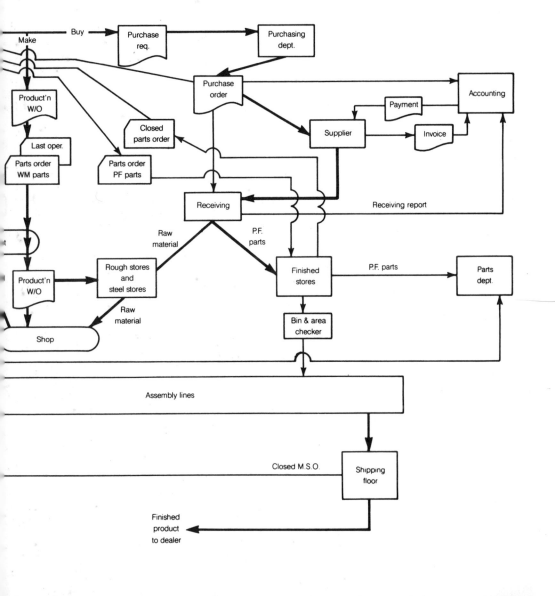

Materials
Management
Handbook

T. H. Allegri

Notices

dBASE®	Ashton-Tate
IBM®, **PC**®, **XT**™, and **AT**™	International Business Machines Corp.
Lotus 1 – 2 – 3®	Lotus Development Corp.
Micro-MAX MRP™	Micro-MRP, Inc.

FIRST EDITION
FIRST PRINTING

© 1991 by **McGraw-Hill, Inc.**

Library of Congress Cataloging-in-Publication Data

Allegri, Theodore H. (Theodore Henry), 1920 –
    Materials management handbook / by T.H. Allegri.
      p.   cm.
    Includes index.
    ISBN 0-8306-3513-0
    1. Materials management.   I. Title.
    TS161.A37   1991
    658.7—dc20   91-13904
                            CIP

For information about other McGraw-Hill materials, call 1-800-2-MCGRAW in the U.S. In other countries call your nearest McGraw-Hill office.

Vice President & Editorial Director: Larry Hager
Book Editor: Suzanne L. Cheatle
Production: Katherine G. Brown
Book Design: Jaclyn J. Boone   EL2

Contents

Introduction

Materials management is a concept that is congruent with the modern view of manufacturing because it organizes, coordinates, and integrates the functions of procurement with the logistics of manufacturing operations. Advances in computer technology have given this concept a further impetus, since it is now possible to obtain real-time operating data to fine-tune the entire manufacturing logistics program.

The plant departments that are most generally involved in materials management operations are purchasing, traffic, production control, materials handling, material scheduling, and inventory management; however, the implications of the concept go far beyond these departments in the manufacturing entity. For example, there is additional involvement with manufacturing and plant engineering, maintenance, engineering design, finance, and accounting. In a word, materials management, if carried out to its ultimate objectives, requires the complete coordination and cooperation of every manufacturing-plant department.

Like many complex systems, however, materials management has the fortuitous capability of being undertaken in gradual stages. So although the ultimate goal might be total plant involvement, it is possible to phase in one step at a time so that the shock of introducing the concept becomes so innocuous that most of the "players" (employees) might not even be aware of the cumulative effects of the materials management program until operating statistics produce the revelation that performance and productivity have been improved significantly.

Another important aspect of the materials management concept is that small segments of the program can be used profitably without the necessity of undergoing radical upheaval in manufacturing. Management might subscribe to only a small segment of the overall philosophy of materials management and that might work to their benefit since other problems and complexities of their operations might be limiting factors. Some of these difficulties might be the result of labor concerns, capital requirements, product mix, and the like. Nevertheless, some of the features of the materials management concept might be advantageous—for example, the linking of purchasing with inventory management and production control.

This and other combinations of a materials management program that might evolve in certain plant situations will be discussed in this book, as well as a methodology for introducing materials management in the manufacturing environment.

Also, since materials management is a planned integration of many technologies and subsystems, some of the salient subsystems are storage and retrieval, internal and external distribution, information processing, and the efficient use of facilities. In addition, profit and productivity centers, flexible manufacturing and assembly operations, and procurement practices will be examined in order to maximize customer service, minimize inventory investment, and maximize the overall effectiveness of plant operation.

1

Materials management concepts

Introduction to materials management

Production systems and equipment are advancing the state of the art in manufacturing in an unprecedented manner as we enter into the twenty-first century. This rapid rate of change in methods and hardware has affected virtually all aspects of the factory. Thus in the last decade or so, the term *materials management* has surfaced in the vast and complex infrastructure of the contemporary plant.

Precipitous rises in materials costs and the no-nonsense view of scaling down large inventories have caused many companies to reevaluate the overall flow of materials from suppliers to finished product and ultimately to the consumer or user. Based upon these evaluations and their conclusions, many companies have created a new position, that of materials manager, with a range of success from mediocre to excellent. The index of success with the reorganizations of materials departments and plant structure has been the profitability of the enterprises concerned. Successful implementation of the materials manager concept has, in many instances, exceeded expectations; therefore, one of the main purposes of this book is to demonstrate how to achieve the best possible application of this concept to suit the needs and ambitions of your organization.

The task of obtaining better materials flow is not a simple one because a complex combination of forces, both internal and external to the corporate entity, must be taken into consideration before the departmental functions can be organized to make the concept work, either in whole or in part.

It is no small wonder that in the last decade sharp rises in materials' costs have given concern to inventory buildups. The buying, inventorying (storage), and shipping costs usually represent 30 to 50 percent of the *product cost*, the true cost of materials. Since so much value is a characteristic of materials, some of the shrewder company managements have come to the logical conclusion that they had better use professional managers to resolve cost and delivery problems and to upgrade the management of the flow of materials into, through, and out of the complex web of manufacturing.

1

The solution to the problem of the quality of flow, or *flow management*—or materials management, as it has been called—has long been debated in corporate offices. Resolving the problem has led to an increase in the number of materials managers (MMs), whose titles have ranged from the glorified MM to Logistics Manager, Physical Distribution Manager, Planning Manager, Supply Manager, and possibly some other more or less descriptive honorifics; and, departmentally, MM has been called Logistics Management and Logistics Engineering. Another discussion, besides the all-encompassing one as to whether or not to adopt the MM concept at all, is whether the materials management function belongs on the divisional or the corporate level. Let me just say, for the time being, that the MM approach might not be suitable for all companies, but it must, in every instance, be tailored to meet the needs of a particular company.

In the classical, theoretical version of the MM concept, MM supervises the functions of purchasing, production planning, production scheduling, inventory control, and distribution, including the traffic functions. In more traditional companies, purchasing reports to the president, production planning to manufacturing, distribution to marketing, and traffic to either marketing or manufacturing. If MM is to succeed, however, it must have very high visibility in the company, so the MM should report either to the president or general manager, or at the very least, to the manufacturing vice-president.

It is all well and good to discuss the visibility and functions of the materials manager, but something further should be said for the compelling reasons why MM as a concept should be adopted wherever the proper climate is generated for it. For example, if a company finds that a sizable portion of its funds are being consumed by material and transportation costs, or if inventories are getting out of hand because of overages, or if critical shortages are hampering production, then that company is a candidate for the MM concept.

When material and transportation costs approach 50 to 60 percent of sales, then by decreasing costs only 5 percent, more than 10 percent of sales would have to be made to generate only half the amount of profit, or 5 percent. Thus, the materials and transportation cost avoidance of 5 percent is tantamount to adding 10 percent to the annual sales figures—a worthy goal in any company! Also, in many companies, there are times when there is a necessity for an inventory buildup; an example is a "strike hedge," when a large supplier, usually of primary goods, is facing an impending union contract negotiation and the "buying" company does not want to be caught short if the contract cannot be negotiated readily and the supplier goes out on strike.

In general, unless there are such exigencies, a company is well served to keep its inventories lean and trimmed. As a result, the company must require better response time from suppliers and perform well regarding customer deliveries. To achieve this fine balance, then, requires keener attention to materials flow, ergo the necessity for MM. From a financial and tax standpoint, also, lower inventories usually assist the cash flow, and this type of cost avoidance aids the appearance of the balance sheet.

Traditionally, when costs get out of hand, each of the many plant departments is given strict instructions to lower costs. This plan might work in some instances, for short periods of time, where the purchasing group is told to lower material costs; production and inventory control divisions are instructed to lower inventory levels; manufacturing is told to improve productivity; and distribution is badgered to improve customer service. In the long term, however, it has been found that since all of these departments and divisions of the company are more or less related into a complex system, as it were, this sporadic push to control costs in the department makes another department ineffective, especially when "out-of-stock" notices are sent to manufacturing and crucial component shortages militate against effective customer delivery schedules!

So in order to avoid looking at the problem as a mix of isolated parts, it was a logical conclusion for the top echelons of company administration to attack the difficulties as a coordinated whole and to integrate the functions of materials movement into the materials management philosophy, which requires the shift of control relationships from the individual units of production and inventory control, purchasing, and distribution, et al., under one span of control so that reporting is channeled to and through the materials manager. Thus, instead of separate activities vying against each other, the MM has the capability of controlling the flow of materials and product from the sources of supply (vendors), through the manufacturing operations, into storage when required, or out through the customary channels of distribution to the customer and/or user.

Advocates of the MM philosophy can easily justify their thesis that a materials manager eliminates the conflicts that arise because of the vested interests that each purchasing, production control, and inventory control department has when it is a separate entity. Under one materials manager they are required to work together as a team in a concerted effort to improve the flow of materials in a finely orchestrated manner. In this way, a professional materials manager enables a company to obtain the cost, inventory reduction, and quality, as well as performance, improvements that would not be possible if each functional area remained the guardian of its own turf.

Despite the fact that the cost of purchases, inventory levels, and delivery schedules are among the factors that determine how profitable a business enterprise might be, these quantities are mutually related and lend themselves to single, managerial control. The tighter this control is, the better and more profitable the company will become, all other factors remaining equal. As an example, inventory levels depend not only on the way in which materials are purchased but also on the level of customer service and the delivery times involved. Also, when one supplier is higher in cost but more dependable than a low-cost supplier, material costs must be balanced against manufacturing costs and delivery schedules. Therefore, a professional materials manager can weigh the trade-offs and make decisions in these instances, where, in the traditional monolithic structure, these concerns might not be as obvious and solutions might remain clouded over until irreparable harm has occurred; i.e., the loss of customers.

Materials management organizational structures

In the traditional monolithic organization in which purchasing reports to the president, production control and inventory control report to manufacturing, etc., the opportunities for blocking sound material flow are legend.

Figure 1-1 illustrates the type of vertical communications flow in the traditional organizational structure that actually militates against high productivity and the least total cost of product because of the conflicts arising through self-interest. Most company organizational charts are not as simplistic as represented in the figure, and that applies also to the description of materials management structures that have been

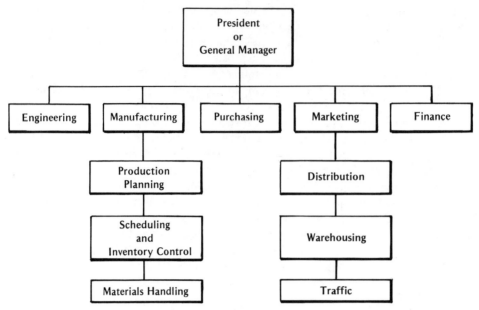

Fig. 1-1. The traditional organizational structure for a manufacturing company.

depicted. The illustrations have been purged of extraneous elements so that the relationships they represent will be obvious.

Figure 1-2 illustrates the basic MM structure. This completely integrated MM structure is said to represent approximately one-third of all the companies that have adopted the MM philosophy. This is, however, only one of several forms of organizational structure representing the materials management concept. In every MM organization there is usually at least one professional materials manager, customarily bearing a title such as Materials Manager, Logistics Manager, Physical Distribution Manager, or the like. To be considered an MM professional or specialist, the manager has to have responsibility for several of the materials functions of purchasing, production planning, scheduling, production control, inventory control, distribution, and materials handling.

There are several other forms of MM organization. The MM organization can report to the supply side, distribution area, or manufacturing, since there is an inherent flexibility to the concept, which is based primarily on the flow and movement of materials through the plant, in addition to considerations of transportation to the factory for the initial stages of production and to the customer or user at the end of the distribution chain.

Supply-sided materials management organization   Figure 1-3 illustrates a partially integrated organizational structure in which purchasing, production planning/control, and inventory control report to the MM. Notice that the structure has been tilted to the supply side because this particular type of company is closer to the source of its supplies and vendors.

Distribution-type materials management organization   Figure 1-4 shows another partially integrated MM organizational structure, in which distribution, traffic, materials handling, production planning/control, and inventory control all

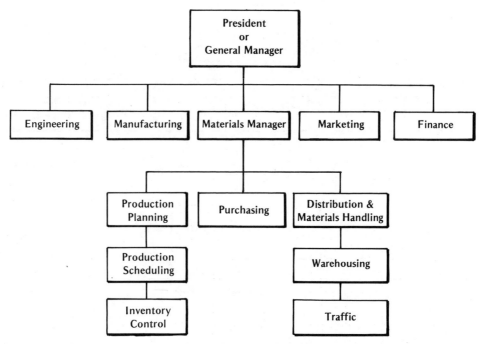

Fig. 1-2. The basic, completely integrated, materials management organizational structure for manufacturing.

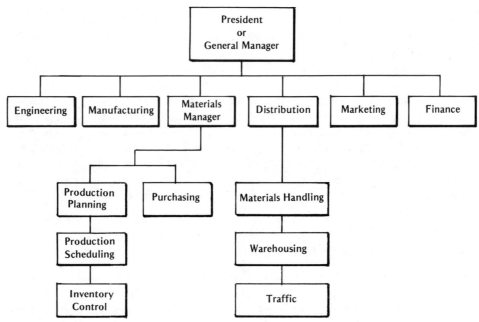

Fig. 1-3. A partially integrated materials management structure that emphasizes the source of supply.

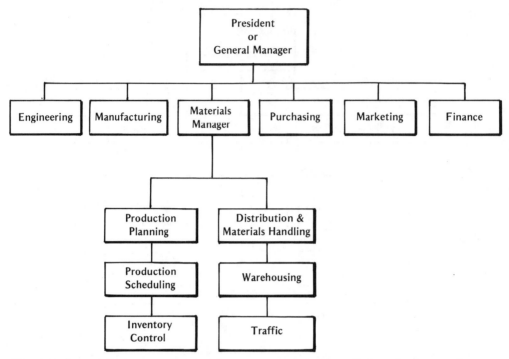

Fig. 1-4. A partially integrated materials management structure that emphasizes market areas.

report to the MM. In this type of organizational grouping, the materials movement relationship indicates that the company plant(s) might be located closer to their markets than to their sources of supply.

Manufacturing-type materials management   Since manufacturing is in the middle of the flow of materials from source to user, that is probably the most logical location for the materials manager and his interrelated activities. The MM functional organization reporting directly to the head of the manufacturing group can be responsive immediately to conditions on all of the production lines, operations, and processes so that down-time due to materials shortages can be held to a minimum, if not eliminated completely; thus manufacturing departments can virtually eliminate "short-sheets." Using the methodology of Just-in-Time manufacturing, inventory effectiveness can be enhanced by all of the fine-tuning that computerized shop floor control will permit. (See FIG. 1-5.)

Materials management paradigms

In many companies, observers have found that organizational charts—even complex, all-encompassing structural denominations—do not often reflect the real, effective chain of command. Material relationships cutting across company lines often come about because of the leadership abilities (usually) or just good, old common sense of one of the more experienced employees. This is the person in the company who many

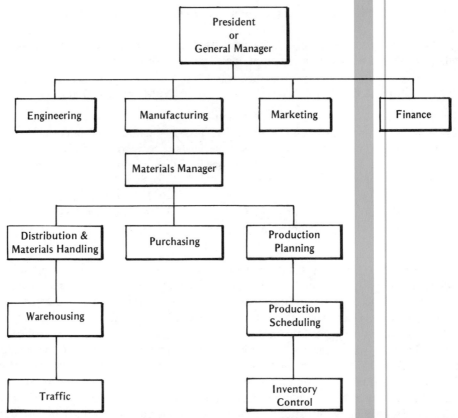

Fig. 1-5. A materials management structure that emphasizes the manufacturing side of business.

employees approach when new instructions have to be interpreted, when new rules are passed down to the rank and file, or when union contract negotiations are in process and various labor-management articulations have to be digested. The daily worth of this company savant, however, is usually demonstrated when some departmental crisis has to be resolved; when various parts or components are in short supply and red tape has to be cut in order to make specific, tight delivery schedules; or when somebody just does something that is completely asinine and a quick recovery has to be accomplished in order to save the company money. For example, sometimes trade-offs have to be made with suppliers in order to obtain quality levels that should have been included in, but were inadvertently omitted from, the purchasing specifications.

Although lateral relationships enable some companies to function smoothly most of the time by moving the level of decision making down to the point where the information exists, rather than bringing the information up through channels to the points of decision, the volume of these decisions, handled in this informal manner, can by their very nature be limited in quantity—and are the exception rather than the general rule. Thus, to depend on this type of communication network when the volume of transactions increases is to court disaster. It is this realization—i.e., of the shortcomings of the traditional organization—that has fostered the development of the materials management philosophy.

Some of the trade-offs, or relationships, that have to be taken into consideration when contemplating the MM organizational structure are as follows:

Purchasing	Inventory Control
Price vs. quality	Inventory costs vs. price
Price vs. inventory levels	Inventory costs vs. manufacturing costs
Price vs. transportation costs	Inventory costs vs. transportation costs
	Inventory levels vs. warehousing (storage) costs

Distribution	How Far to Go with
Transportation costs vs. inventory levels	Lateral Relations?
Transportation costs vs. customer service levels	

As indicated previously, many companies have used informal means to communicate across departmental lines in the traditional organizational structure. In many instances the lines of communication that are not shown on the company's organization chart have been used to solve crises that are, basically, materials problems. Unfortunately, companies having to rely on these informal channels are reacting to problems, rather than anticipating them. This methodology has led many top officials to consider the possibilities inherent in the materials management concept. The conclusion is that it is more economical and profitable to avoid the problems (usually associated with materials movement in the traditional form of organization), rather than react to them, when the responsibility for materials movement is a shared one, rather than vested in one manager with cognizance over most, if not all, phases of materials movement.

Computers, bar coding, and other automatic identification techniques

Programmable controllers and computers have established themselves in virtually every facet of manufacturing, especially where materials movement is concerned. The exceptions are the very smallest of the "alley shops," where no modern data collection or information systems are used. With the cost of PLCs (programmable logical controllers) and computers decreasing at a precipitous rate, even some of these shops might eventually take advantage of the effectiveness to be derived from their use; however, smaller, marginally profitable firms might question the time and money required to be invested in the hardware and software systems that are needed.

From directing labor and controlling equipment on the shop floor or in the warehouse, computers and PCs have become a formidable asset to manufacturing operations.[1] When the computer is integrated with bar coding or other means of automatic identification, the power to control operations is further enhanced.

Since the very foundations of a computer operating system are the reliability and timeliness of the data that drives the system, it is still quite remarkable that many of today's companies that have computer systems rely on keyboards for data entry instead of the vastly superior, relatively error-free, bar code data entry methodology. Bar coding, however, is one of the fastest growing data entry methods in use today, comprising approximately 80 percent of installations using automatic identification means, followed by magnetic strip data collection, which is about twice in number of installations as either optical character recognition (OCR) or radio frequency (RF) technologies. A small number of companies, slightly over 2 percent of auto. ident. installations, use voice recognition and surprisingly, three times that number use machine vision.[2]

As an indication of areas within the manufacturing and distribution complex where automatic identification, primarily bar coding, is finding its largest usage, the following list will give you some idea of this proliferation:

- Incoming material storage
- Processing operations
- Time and labor reporting
- Work-in-process storage
- Assembly operations
- Quality control
- Finished product storage
- Order picking and sorting
- Receiving
- Shipping

Using good data-entry techniques such as automatic identification methodologies, primarily bar coding, many companies are using their computers to support planning and scheduling activities. A large number use computers for Materials Requirements Planning (MRP), Distribution Requirements Planning (DRP), and shop floor control. These topics will be discussed in greater detail in a subsequent portion of this text. DRP is an offshoot of MRP and includes the requirements of the distribution system in the planning of production and inventory levels.

MRPII, or Manufacturing Resource Planning software is often run on mainframe computers; however, minicomputers and personal computers are becoming more widely used for these tasks. Also, some manufacturers might use more than one type of computer for MRPII functions. As an example, a *mainframe*, or host computer, might hold the central database while one or more PCs generates reports, performs analyses, and the like. Some of the latest PC versions are powerful enough on their own to perform host computer functions in many companies.

Many tasks performed by a professional materials manager can be performed, at least partially, by computer-based systems and procedures. This fact is particularly evident in smooth-running, well-established organizations, in which specific materials movement functions, such as the reduction of production plans into material requirements, can be entrusted to a checklist of fixed decision rules and data transmissions that a computer handles readily and with better precision and speed than in any other methodology. It is this primary reliance on the computer that has led some companies to forego the employment of professional materials managers in their organizational setups. On the other hand, a number of companies use materials managers to implement computer-based systems for controlling materials flow.

Companies that attempt to substitute computers for materials managers are leaving themselves wide open for future trouble, in that there is no adequate substitution for a professional materials manager, who is the focal point for decision making. All too often in the movement of materials, where several departments are involved, the resolution of conflicts and the problems of turf militate against the achievement of the least total cost because profit centers, or their equivalent, are involved. The materials manager to whom all of these groups would report can forestall, and in general, eliminate problems of turf, which hurt rather than help the profitable use of a company's resources. In general, computer-based materials systems cannot safely supplant the materials manager organizational concept in the long haul.

Organizational design has been studied and evaluated for decades. There is nothing simple about the subject because personalities, character traits, available skills, personnel considerations, and a host of other factors influence decisions involving a materials organization. The primary concerns for an organizational structure, however, should be the work to be done and the trade-offs that must be made.[3]

Current trends

While many companies are reorganizing their procurement and logistics functions in the materials management mode, there are also a number of changes being made in purchasing practices. For example, a major emphasis is being placed on centralized buying and commodity divisions. The multinational company is also undergoing change in these areas as worldwide sourcing of materials is made more cost-effective through the dramatic increases in efficiency of transportation methods.

National procurement operations   Companies that have plants and warehouses in a number of different locations within a state or in several states have found that they might have several buyers in scattered locations who are buying the same commodities. There are several disadvantages in this methodology. As an example, a company that is a heavy steel purchaser might find out that its buyers in different plant locations are bidding against each other; in other words too many buyers are chasing too few products, with the suppliers taking advantage of the users in this instance. When centralized and coordinated purchasing programs are established, one of the primary advantages, of course, becomes the realization of lower prices as a result of the larger volumes that can be purchased. Another benefit, which is brought about by the larger dollar volume, is better delivery schedules. There is also the possibility that better and more consistent quality levels can be achieved, together with greater measures of standardization.

If purchasing functions can be centralized at a single location, then the buyers and their supporting personnel can free up personnel for other duties, or their numbers can be reduced through attrition. In the main, centralized, specialized procurement operations are cost-effective from the standpoint of office space, personnel, and equipment as the duplication of these things is eliminated or reduced.

International procurement   Multinational companies are not the only purchasers of goods made offshore. Although overseas purchasing is a great deal more complex than stateside purchasing, more and more companies have found it profitable and advantageous. Because of the restrictions imposed on trade and the tariffs imposed by most countries on imported articles, certain percentages of a finished product must be manufactured in the country of importation. For example, prior to the most recent U.S.-Canadian accords on tariff for certain articles produced in Canada for exportation into the U.S. the product would have had to contain at least 35 percent of the value of the product produced in the United States, thus minimizing the duty levied against the article.

Ocean transportation with containers of 20-, 25-, and 40-ton capacity and the like is a rapidly increasing method of shipping goods between countries. Another impetus that transoceanic traffic has received has been derived from the arrival of the wide-bodied aircraft on the transportation scene. Millions of pounds of freight are in the air at any given moment of the day or night. So, between the intermodal container that travels by boat, train, over-the-road carrier, and air freight, overseas procurement is limited only by the imagination of the entrepreneur and the freight forwarder.

Summary

Materials management is a concept that is congruent with the modern view of manufacturing because it organizes, coordinates, and integrates the functions of procurement with the logistics of manufacturing operations. Advances in computer technology have given this concept a further impetus since it is now possible to obtain real-time operating data to fine-tune the entire manufacturing logistics program. The plant departments that are most generally involved in materials management operations are purchasing, traffic, production control, materials handling, material scheduling, and inventory management; however, the implications of the concept go far beyond this in the manufacturing entity. For example, there is additional involvement with manufacturing and plant engineering, maintenance, engineering design, finance, and accounting. In a word, materials management, if carried to its ultimate objective, requires the complete coordination and cooperation of every manufacturing plant department.

Like many complex systems, however, materials management has the fortuitous capability of being undertaken in gradual stages so that, while the ultimate goal might be total plant involvement, it is possible to phase in one step at a time so that the shock of introducing the concept becomes so innocuous that most of the "players" (employees) might not even be aware of the cumulative effects of the materials management program until operating statistics reveal that performance and productivity have been significantly improved.

Another important aspect of the materials management concept is that small segments of the program can be used profitably without the need to undergo radical upheavals in manufacturing. The management view could be that they will subscribe to only a small segment of the overall philosophy of materials management, and that might work to their benefit since other problems and complexities of their operations could be limiting factors. Some of these difficulties might be the result of labor concerns, battles over turf, capital requirements, product mix, and the like. Nevertheless, a number of features of the materials management concept might be advantageous—for example, the linking of purchasing with inventory management and production control or distribution with materials handling, and so forth.

Also, since materials management is a planned integration of many technologies and subsystems, it would, by its very nature, include storage and retrieval (in warehousing and inventory-related areas), internal and external distribution, information processing, and the efficient use of all the plant's facilities and resources.

Although the efficient use of all of a company's resources that are available is a worthwhile and laudable goal, it is so broad in its general terms that we must address more specifically the area of profit and productivity centers, flexible manufacturing and assembly operations, and procurement practices in order to maximize customer service, minimize inventory investment, and maximize plant operating effectiveness. Other areas that are influenced and affected, and that must make contributions to the materials management concept, are as follows:

- Distribution resource planning—DRP
- Material requirements planning—MRP
- Manufacturing resource planning—MRP II
- Capacity requirements planning—CRP
- Manufacturing computer systems
- Manufacturing decision-support systems

Manufacturing interfaces with the following areas:

- Marketing and forecasting
- Quality assurance
- Finance
- Purchasing
- Engineering
- Distribution and warehousing
- Transportation
- Master scheduling
- Shop floor control
- Production planning
- Inventory management

Notes

1. T. H. Allegri, *Advanced Manufacturing Technology* (Blue Ridge Summit, PA:TAB Books, 1989).

2. Op. cit.

3. Jay Galbraith, The European Institute for Advanced Studies in Management, *Designing Complex Organizations* (London: Addison-Wesley, 1973).

2

Marketing, forecasting, and management science

An overview of marketing and forecasting

Determining what the company shall produce—i.e., the nature of the product, size, shape, etc.—is essentially a function of the marketing department. Such things as the color of the product, the logo to use, and so forth, are also concomitant interests of this group. Another function that is usually associated with marketing is the very important question of "How many?" In terms of producing a number that is realistic for each product—by model, by series, by size, and the like—there is an arcane science performed by marketing that is called *forecasting*. Many a marketing head has rolled when the number of products manufactured has vastly exceeded demand. Contrary to logical assumption, however, very few heads have been severed when the reverse is true.

The old adage that, "Nothing happens until somebody sells something," presents another side of the marketing picture that is closely related to sales and distribution. The feedback from the sales department provides a good deal of the data that is factored into marketing forecasts, product pricing, styles, and a host of other variables inherent in the numerous models of a product, or product lines, that are the output of the modern manufacturing enterprise.

In general, the beginning point of the forecast is a prognostication of economic conditions for the coming year. Of course there are longer time frames and projections based on several years—five years, ten, etc.—but because these projections have to be updated annually, this discussion centers around the one-year forecast, which is about as realistic as it is possible to extrapolate because the materials management department's budgets are usually focused on this time period and this functional department.

The next step in forecasting is to determine the industry's total market potential for the product, then to assess the share of the market that the company will, or should, obtain. Thus, by applying the measures of total market to market share, the projected sales volume for the coming year becomes the target figure upon which the operating budgets of the company are based.

The implications of marketing decisions

In today's socio-economic structure and environment, the business enterprise, and with it the manufacturing complex, is dependent to a large extent on the results of marketing decisions. That is to say, almost all of the planning, organizing, and creative activities associated with the production of goods and services—the "putting into iron" of manufacturing (and business)—is better served by the increasingly effective marketing effort.

By and large, marketing decisions are reflected in materials management functions because the amount of investment in inventory is directly related to marketing data and indirectly related to the flow and movement of materials to meet delivery schedules. Optimizing the flow of materials, which is the single most significant aspect of the MM rationale, contributes broadly to the maximization of plant operating effectiveness. Improvements in plant operations, then, make it possible to provide better customer service—and, customer satisfaction leads to increased sales and profitability.

In order to have a better understanding of the materials management philosophy, apart from its methodology, you must consider the significance of various factors in marketing decisions that can make or break the company. It has been noted that external forces such as economic conditions, government policies, competition, and global affairs impinge on marketing decisions and will have an effect on the sales of a company's product. For example, a multinational company with a substantial overseas market could be disastrously affected by a change in the value of currency either at home or abroad. A lack of diversification in a company's product line will have a domino effect in such a situation because too many of the company's eggs are in the same basket. Also, the value of money, i.e., interest rates—will tend to increase or decrease the plant's ability to compete as the rates rise or fall.

Marketing decisions, then, are not static and must be monitored continuously to fine-tune the enterprise. The feedback from marketing through the planning activities of the company will directly affect the functions of materials management as manufacturing operations are scaled up or down to coincide, as closely as possible, to the external forces that are always at work.

The tools of management science

Because materials management is a budding science, it is only fitting that a brief history of its basic origins be traced here in order to illustrate how it has developed and prospered in the realm of enlightened management. A better understanding of its formative roots will enable the practitioner to adapt its various complexities into the very fabric of his/her own company's structure with a greater degree of freedom than would otherwise be the case.

The origins of operations research

There are many ways to describe management science. A few that come to mind are: operations research, operations analysis, and systems analysis. Fundamentally, these descriptions all boil down and can be synthesized as a systems approach to the solution of problems, events, and the like that pertain, especially, to the management of an enterprise.

Engineers and scientists—there is a major differentiation between the two: one applies what the other discovers—have emphasized the importance of employing mathematical modeling to solve complex problems. During World War II, a number of military problems, mainly in the areas of logistics—for example, troop, equipment, and supply movements—were resolved through the use of modeling. After the war, operations research (OR) methods were applied to production problems in some of the larger companies. With a number of successful applications behind it, therefore, OR began to be treated as a management tool.

Criteria for operations research

Although OR is not for everybody, certain criteria must be followed to use it to the fullest extent of its effectiveness. Experience has shown that this methodology can be a very formidable tool when it is properly applied. In the first instance, to be effective OR must have the sanction of the company's top management layer because the result of its application requires management follow-through and direct action (and, sometimes, direct intervention).

Another element that requires consideration is that comparisons of several feasible alternative actions should be based on measurable values that are related to such specifics as costs, sales volume, and rates of return on investment (ROI). Other specifics—e.g., measurable quantities—can be substituted for the ones described; however, the basic premise is that the quantity must be measurable and distinct.

An essential element, also, is that the mathematical model upon which the solution is based should be so well defined and explicit that any other modeler (analyst) can take the data to be manipulated and achieve the same results.

Last is the question of programming and computer time. In many mathematical modeling problems, the quantity of data to be processed usually exceeds the ability of manpower alone to process the data manually; thus, computer time is essential. This requires a commitment on the part of management to supply not only the computer hardware needed to solve the problem, but also the programming time to formulate the data for computer consumption. Fortunately for the management sciences, the computer has developed to a level of sophistication in which it is not even necessary, in most instances, to use a mainframe computer for most mathematical modeling. The vastly enhanced power of minicomputers, micros, and even PCs is sufficient for most modeling programs.

The classification of OR models

It has been indicated in the foregoing paragraphs that mathematical modeling is one of the characteristics of management science, with important implications for materials management. The reason it is deemed significant is that it offers the materials manager the opportunity of testing hypothesis without performing a live demonstration of a particular event or series of events. In fact, it is not often possible to perform a materials flow experiment in a manufacturing environment to test the practicality of a method. With the use of mathematical modeling, if the experiment works, then everything is just fine. But suppose, because of some unforeseen glitches or physical impracticalities, the experiment ends in failure? Then the credibility of the MM department is strained, and possibly there is an awkward mess left in the factory requiring many labor-hours and dollars in overtime to repair the damage.

As an example, suppose the materials manager would like to determine inventory levels to achieve the lowest total cost when considering purchasing costs, transportation, handling costs, and storage charges. The problem becomes somewhat more sophisticated than the academic Economical Ordering Quantity (EOQ) equation was ever meant to be. The parameters are enlarged because of the extraneous factors that have been introduced. Merely factoring in the transportation costs based on the geographic location of different supply sources causes this problem to require a good portion of computer time. Nevertheless, with the development of a correct mathematical model, the MM can obtain a number of plausible solutions to the inventory problem. This simulation of a real-life situation makes this management tool a worthwhile addition to the MM department's kit bag.

Another advantage to modeling is rather subtle and comes about indirectly. As the problem to be simulated is studied in order to formulate the equation or series of equations for the model, so many questions of the "What if?" type are asked that everyone associated with the problem obtains a better grasp of the subject. Therefore, the modeling represents a learning experience that is very effective and beneficial for all the personnel involved.

There are several different types of models that might be of value to the MM staff. Essentially, mathematical models can be classified by their purpose.

A model may be said to be *descriptive* if it describes how a system works—i.e., it describes things the way they really are and makes no value judgments concerning the specific matter being studied. The descriptive model does not select the best possible solution; it merely represents the event more or less as it is. It can display the various possibilities that are available, and it might point the way for the observer (user) to decide what the consequences of each alternative solution might be.

Normative, or *prescriptive*, models include predetermined criteria that indicate how the system should work in order to achieve a specific objective. It is this capability of optimizing a solution that has given this type of model the name *optimizing model* or *decision model*.

Another type of model is called *deterministic* when there are no probabilistic considerations—in other words, when the laws of chance do not come into play. As an example, the equation $I = PRT$ (Interest = Principal × Rate × Time) is considered to be deterministic because all the factors are fixed. There are no variables because all factors are precise quantities, and the solution to the equation is determined by these exact relationships. Furthermore, there is no uncertainty in the conditions pertaining to the equation.

On the other hand, once conditions of uncertainty are introduced into a model, it is said to be *probabilistic*. The underlying consideration of random conditions and uncertainty requires the mathematical reasoning of statistics as a basis for the formulation of the model. Actuarial tables used in pension plans, such as social security, or life expectancy tables formulated by insurance companies are based on probabilistic models.

Decision theory

In the previous section, it was indicated that mathematical models might be either descriptive or normative, and be composed of deterministic or probabilistic variables. To round out this discussion of modeling, a few words will be devoted to what has become known as *decision theory*.

Decision theory is based on a foundation of statistics and the behavioral sciences, and attempts to provide a systematic analysis for decision making. In materials management there are a number of instances in which decision theory can be, and should be, employed, as we shall see later in this section. One of decision theory's primary tenets is to take decision making out of the realm of art and make it more of a science. In order to do so, it gathers all the particular elements generally associated with decisions and thereby provides a methodology for the analysis of a complex problem containing a number of alternative solutions with probable results. In this manner, decision-theory models are normative in purpose, while containing probabilistic variables.

Decision theory modeling has implications for MM in the following areas:

Inventory model In the manufacturing enterprise, one of the most vexing problems encountered is the business of balancing inventories between having stock and being out of stock. There are dangers and disadvantages in both situations because, when the plant has too much inventory on hand, there are carrying costs to consider—i.e., the cost of storage and the use of needed space—as well as the dangers of obsolescence and shelf life. If there is not enough inventory to finish assembling a product on the factory floor, and if there is a stock-out problem, production lines might be shut down or diverted to other production, with a loss of productivity as a result. With a tight supplier network and an entirely different philosophy, the Japanese have resorted to the principle of Just-in-Time (JIT) manufacturing to solve this problem.

In JIT manufacturing as it is practiced in Japan, it is customary for the supplier to provide quality parts to point(s) of the production line in the desired quantities according to a prearranged schedule. In some Japanese plants, the parts might be transported by the supplier to a particular door of the factory, rather than brought up to the production line. As you can imagine, the cadre of reliable suppliers becomes the very muscle and sinew for Japanese manufacturing productivity.

It is difficult to envision such fine-tuning of inventory in many U.S. companies, with the exception of such large-scale manufacturing entities as the automotive and appliance segments.

If we can regard JIT manufacturing as the ultimate objective for the control of inventory, then our inventory model must answer two fundamental questions that concern quantity (or, how many?), and timing (or, when are the parts required?). Other parameters of the mathematical model are as follows:

- Cost of storage
- Cost of purchasing
- Cost of transportation
- Cost of containerization
- Cost of packaging
- Economic ordering quantity (EOQ)

- Optimum reorder point
- Other aspects peculiar to the particular plant

Allocation model In all companies there are always a number of programs and projects that compete for available funds. Because there are usually more projects than there are dollars to satisfy these requirements, the decision-making entity has to determine what projects will maximize the return on investment. In addition, the materials manager must determine the optimum manner in which to allocate the scarce resources at his/her disposal.

Since the MM will be competing with other department heads for funds, the method of presentation that must be employed to gain acceptance for the particular proposal should be capable of withstanding rigorous examination. The widespread use of linear programming models bears evidence that the MM would be well served to resort to this methodology to support the position taken.

Linear programming delineates the objective(s) to be achieved in the form of a mathematical function, the value of which is to be maximized (profit) or minimized (cost). In defining linear programming mathematically, we can say that it is the analysis of problems in which a linear function of a number of variables is to be maximized (or minimized) when those variables are subject to a number of restraints in the form of linear inequalities. The Simplex method of linear programming uses a procedure that searches the feasible alternatives in order to find the particular one that maximizes (or minimizes) the value of the function.

Queuing model In the area of materials management, whenever there is a stoppage in the orderly flow of materials, operating costs that are normally avoidable will occur. The mathematical modeling that will help you to understand, and thus to minimize, the waiting line that sometimes occurs when customers arrive at a service area is based on *queuing theory*. The first mention of the application of queuing theory was in the work of Karl Erlang in the field of telephone engineering. Since that time the queuing model has been applied to a number of waiting-line problems that affect MM as well as a host of other disciplines.

Workers waiting for tools at a tool crib, trucks waiting to be unloaded or loaded, and trucks waiting for dock spaces are all problems that lend themselves to analysis by queuing theory because this method of modeling attempts to predict the behavior of the waiting lines so that managerial decisions can be made regarding how much service capacity can be justified economically. There are three significant elements to all queuing models:

1. The pattern of arrivals at the queue
2. The service mechanism
3. The queue priorities

The first element of the queuing model is concerned with the pattern of arrivals of the customers over a period of time. This element serves as input for the analysis. Because the arrivals are at random intervals, the input becomes a *stochastic process*. The second element, which is the service mechanism, describes how the arrivals are handled, or not handled, as the case may be. For example, are there one or several servers? Does each server have a separate waiting line, or do all the arrivals wait in a single line? Because these are nonrandom components of the service mechanism, they can be changed to make the operation more economical or to reduce waiting times.

There are, however, random aspects of the service mechanism; for instance, the service time might vary among customers. Thus the probability distribution of service times must be factored into the queuing model.

Formulating the model permits the behavior of the waiting line to be studied and analyzed. The salient factors might be the length of the waiting line(s), or the amount of time between arrivals and service, and so forth, constituting a number of random variables, rather than fixed quantities.

To mathematically formulate the queuing theory, we have the following:

Let n = the number of units in the service area.

x = the average rate at which the units arrive in the service area.

y = the average rate at which the units are transported out of the service area.

f_n = the probability, or relative frequency, in which there shall be exactly n units in the service area at any given time.

Then,

$$f_n = \left(1 - \frac{x}{y}\right) \left(\frac{x}{y}\right)^n$$

Because this equation describes the long-term, or *steady-state*, condition, if $x > g$ there will be an accumulation of materials (n units) in the service area. Therefore, it is necessary that $a < b$. In the event that $n = 0$, the probability becomes such that the service area will be down for lack of material. In any event, the average amount of material, n, in the service area can be demonstrated to be:

$$A_n = \frac{x}{y - x}$$

This model presupposes that material will move through the manufacturing plant in a random fashion, not according to any fixed schedule. Thus, if the materials manager is satisfied with the random schedule but desires to control the average rate at which material is moved into the service area—in other words, to control the x value—then he/she will be looking at three types of costs for varying the average rate of x, as follows:

1. If it is desired to increase x then it is very likely that more personnel and equipment will be required, thereby increasing transportation (materials handling) costs

2. The MM might want to know what costs might be generated because of a lack of material (which could be the result of poor materials handling techniques or methods of supply; or hangups in prior processing departments)

3. The MM might want to impose in-process storage charges on materials within the service area, based on the average in-process inventory costs, solely for the purpose of cost control.

In review, it is apparent that it might be possible to entertain a value for x that will minimize the cost of materials handling, idle time (down time), and in-process storage because:

1. Materials handling costs increase with x
2. Down time decreases with x
3. Storage costs increase with x

These factors can be mathematically related. Let $M_{(x)}$ = the total cost of a specific methodology for x. Then,

$$M_{(x)} = M_1 x + M_2 P_0 + M_3 \left(\frac{x}{y-x} \right)$$

Where,

M_1 = the materials-handling costs of labor and equipment at an arrival rate x
M_2 = down time cost per unit
M_3 = in-process storage cost per average inventory unit

In order to find the specific methodology that best suits x, the equation can be differentiated, thus providing a value for the mean rate at which material should be moved into the service area.

Simulation model   Simon Ramo, one of the founders of the TRW Corporation, a conglomerate enterprise of multinational reputation, described the applications of the systems approach in each of his two published works, *Centuries of Confusion*, and *Cure for Chaos*. In pursuing the systems approach, Ramo states that it is possible to solve many of our socioeconomic problems, transportation, slum clearance, and assorted dilemmas of our infrastructure through this methodology because, as many engineers will acknowledge, it is possible to cure these pervasive problems only by looking at the whole potpourri objectively and seeing all of the related parts.

Since this thesis is so fraught with possibilities, it would be useful to mention the application of computer simulation as it pertains to problem solving in our smaller macrocosm of the manufacturing world and to treat this complex entity as a system. Computer simulation is one of the most powerful and most widely used methods of analyzing difficult and complicated systems; that is why it will be most useful to materials managers in applying a systems approach to problem solving in the area of materials movement. Also, it is possible to use simulation techniques in at least four areas of concern to the MM staff, as follows:

1. To define a system or a problem within the system
2. To determine crucial elements or components within a system

3. To evaluate possible solutions to complex problems
4. To forecast events and to give direction for project planning

As in the discussion of queuing theory modeling, one of the advantages of computer simulation is that it is possible to design a mathematically logical model of a real-world system without interfering in the real world of the manufacturing entity with all the disruptions that would occur during such experimentation. Therefore, by describing a system in terms acceptable to a computer—that is, by a set of variables, with each set representing a different state or condition of the system—it is possible to substitute many values for these variables to see how they differ from one state to another. When this is done, usually with large quantities of numbers, it is possible to observe the behavior of the model over a period. Depending on the inputs, the output of the modeling will be either deterministic (fixed) or stochastic (random).

Since simulation models are descriptions of a system, the modeler must determine the elements to be included in the model. Thus, in order to make such decisions the purpose for the model should be carefully delineated, and each element to be considered should be evaluated against the purpose as it has been established. The modeler's analysis of each element and the relationship between elements vis-à-vis the purpose have a great deal to do with the results of the modeling.

In general, the model-builder begins with a simple model (or statement) that is subsequently modified as various elements are considered or rejected in the problem-solving process. The various stages usually considered by the modeler are as follows:

1. The purpose of the model is carefully delineated because it is a definition of the problem or system to be analyzed.
2. The system is formulated into a logically derived mathematical relationship in accordance with the purpose of the model.
3. All data pertaining to the model are identified, specified, and collated.
4. The model is now prepared for computer processing with any required data loops, field comparisons, and so forth.
5. As far as it is possible to do so, the validity of the computer program is checked to see that it actually represents the problem as delineated in the purpose of stage 1.
6. A further check of the computer program is made to determine how precisely the simulation model corresponds to the real-world system.
7. The simulation program is run and a series of output values are obtained.
8. After the outputs are obtained, the modeler/analyst reviews the results and then prepares his/her recommendations, caveats, and proposals based on a comparison of the stage 1 purpose with the results of the experimentation.

The final stage of computer simulation is the *execution*—i.e., the putting into practice of the results of the exercise—as outlined in stage 8. It may well be, however, that erroneous assumptions have been made and a revision or modification of the computer program is necessary. In that event, a number of computer programs might have to be run until there is complete unanimity in the effectiveness of the model in representing the real-world system.

Types of modeling   Several types of modeling approaches can be taken by the modeler, depending on how a conceptual framework is conceived.

In a *discrete simulation* model, the dependent system variables change in discrete increments at specified event times in the simulation. In addition, the time variables can be either discrete or continuous, depending on whether discrete changes in the dependent variables are permitted to occur at any time or at specified times. An example of a discrete simulation approach is found in the Queuing Theory model.

In a *continuous simulation*, the dependent variables of the model change continuously over time. In this instance, the model is formulated by establishing equations for a set of variables whose behavior represents, or *simulates*, the real-world system. Models of continuous simulation systems are often described through the use of differential equations, since it is usually easier to formulate relationships regarding the rate of change (dy/dx) of the state variable (the continuous change variable) than to try to describe a relationship for the state variable directly.

As an example of this approach, if we want to show the behavior of the state variable, s, over time, t, at the beginning period of time, o, we would have the following:

$$\frac{ds(t)}{dt} = s^2(t) + t^2$$

Where,

$$s(s_c) = c$$

By integrating ds/dt, we can obtain the response, s, as follows:

$$s(t_2) = s(t_1) + \int_{t_1}^{t_2}\left(\frac{ds}{dt}\right)dt$$

Since a number of continuous simulation computer languages have been developed, there are several numerical integration algorithms that can be found in various tests on numerical analysis, and you are directed to them for further inquiry into this method of analysis.

There are a number of other simulation approaches that depend solely on the view of the real-world situation adopted by the modeler. I will discuss a few of them so that you will become convinced that the horizons for analysis are indeed broad.

In the *event-oriented* approach, the system is modeled by indicating the changes that take place at various event times. The modeler describes the events that might change the state of the system and then develops the logic for each event. Using the simulation system mechanism, it is possible to act out each logic statement in a time-ordered sequence. A number of computer simulation languages have been developed to provide for event scheduling.

Another approach is called *process-oriented*, in which sequences of elements of the simulation model occur in definite patterns, such as in the queuing problem previously mentioned. The logic of such a sequence usually can be included in a single statement.

If you used a simulation language such as IBM's GPSS (General Purpose Simulation System), the statement could then be transformed into the proper sequence of events. The process-oriented language uses such statements to model the stream of events occurring in a system, which are carried out automatically by the simulation language as the movement progresses through the system or process.

Activity scanning is yet another orientation in which the modeler or analyst describes the activities that take place in the system and for which preconditions exist to start and stop the activity. As the simulated time progresses, the conditions for either starting or stopping the activity are scanned. If the preset conditions are fulfilled, the appropriate activity takes place.

To determine whether every appropriate activity has been accounted for, the entire set of activities must be scanned every advance in time. Because of this requirement, this approach is relatively inefficient when compared to discrete simulation and so has not been widely used as a modeling method, despite its suitability for activity durations that are indefinite and determined by the state of the system satisfying preset conditions.

Computer languages Special-purpose computer languages have been developed specifically for the use of simulation modeling, primarily because the technique of computer simulation has become so widespread. One of the major process-oriented languages is IBM's GPSS, and next is Pritsker's Q-CERT (Graphical Evaluation and Review Technique). One of the principal event-oriented languages is SIMSCRIPT, developed by the RAND Corporation, GASP IV and SLAM also have event-oriented capabilities.

Computer simulation is discussed in further depth in any industrial engineering handbook. The "references" section will probably contain mention of A.A.B. Pritsker's book, *The GASP IV Simulation Language*, published by the Wiley company in 1974, and T.J. Schriber's book, *Simulation Using GPSS*, also published by Wiley in 1974.

The purpose of this section was to acquaint you with the possibilities that exist in the realm of management science so you can be more effective in the area of materials management. It is important that the MM be able to approach data processing personnel with at least the proper questions for solving MM problems, leaving the mechanics of computer science to the experts in this field.

3

Finance and accounting relative to materials management

Economic considerations

While the materials management concept has largely found a place in the manufacturing environment, it is not to be supposed that this is the exclusive province of materials management. Quite the contrary, since the materials management concept can be applied profitably to service industries such as truck lines, railroads, and hospitals with equal efficiency.

In addition, lest we attempt to further delimit its application, materials management can be considered as a basic component of any organization or entity that produces a product or service of economic value. When judged in this light, then, we can say that the materials management concept can also be utilized in public and nonprofit organizations to obtain the maximum benefit for expended funds.

Although we like to consider that materials management as a concept is uniquely American, it is like any idea whose time has arrived. It is worldwide in its application and nowhere more appreciated than in Japan, where the Materials Management Society is firmly established with a wide base of support. That the Japanese manufacturing establishment has adopted the materials management philosophy so enthusiastically should give us a moment of pause and introspection.

Since the importance of manufacturing in the United States is declining somewhat and services, both public and private, are accounting for an increasing share of our national output, it is fortuitous that the service industries, per se, are broadly endorsing materials management as a means for increasing the cost-effectiveness of their operations. But, whether for manufacturing or services, the logic of materials management is inescapable, and you will develop the tools for establishing this concept by reading this text.

The ebb and flow of corporate dollars

If you dissect the workings of any manufacturing organization, you will find that most of the cost of the product (i.e., the output) is expended on purchased parts, raw materials, and supplies. Thus, every manufacturer is dependent upon suppliers and is in the process of adding value to the materials that it receives from outside the company.

Figure 3-1 shows, graphically, how the economic balance in manufacturing is achieved. Since half of the company's dollar is spent on purchased materials, the efficacy with which this amount of money is expended should become a vital concern of management. In large, multinational companies we're looking at billions of dollars; in smaller companies the sums of money are smaller, but no less critical. As a matter of fact, the smaller the company, the more there is the necessity of coordinating all of the components that comprise the movement and management of materials because the margins of profit do not allow for many mistakes to be made.

Another hard and uncompromising factor in the movement, storage, or distribution of materials is that every operation adds an element of cost to the total. In other words, the less movement or handling there is, the least cost is involved, except for the storage of the product, where the longer the storage period, the greater the cost involved.

Storage is where the concept of product turnover becomes significant, and annual stock turns of only 1 to 6 times are relatively less profitable than those that are higher. By increasing the number of stock turns, in most instances, the storage cost of the product and the interest on inventory capital can be minimized.

Also, merchandise that is subject to spoilage and pilferage contributes to increased costs because these charges must be factored into the price or cost of the product. This escalation in cost is enhanced when the chain of distribution is fairly complex. In the distribution of a product, each time possession is passed along, commissions and profits must be paid. For example, a manufacturer sells merchandise to a jobber, who in turn sells the product to a dealer. The dealer may sell directly to the retail trade, or he/she may sell to a distributor, or to a manufacturer's representative, who will then sell the product at retail.

As evidenced by these statements, it is apparent that manufacturing and distribution bear a definite relationship wherein one cannot thrive without the other. In both areas, also, it will be found that capital requirements exist. For instance, in manufacturing there is a requirement for a substantial investment in plant, machinery, and in-process inventories. In distribution, the requirements for capital might involve the cost of the goods and the warehouse or stores from which the goods are sold.

Naturally, the profit that is earned on sales represents the return this capital must earn in order to be considered productive. Investment capital that does not earn profits and/or contribute to profitability is considered to be ineffective and requires different orientation and other objective reinvestment. Therefore, one of the major thrusts of materials management is to properly identify inventory that is not working to add value by distribution, and to eliminate it.

Analyzing assets

If we were to study a typical manufacturing company, we would find that it uses most of its capital to add value to a product by manufacture, although a substantial part of its

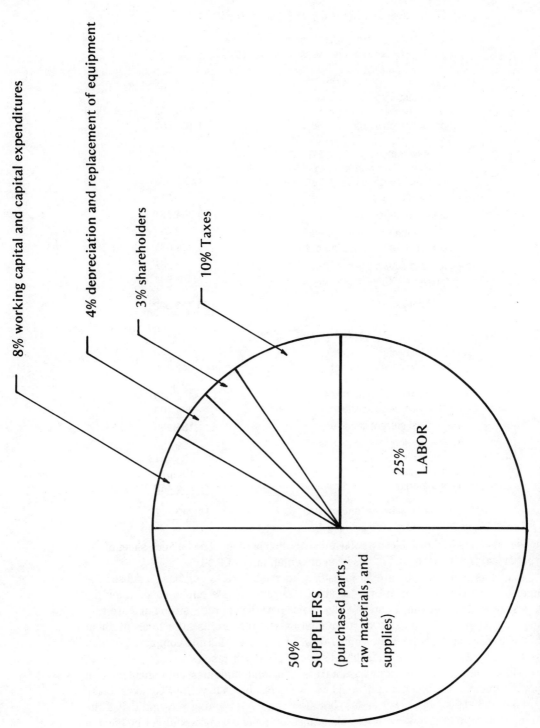

8% working capital and capital expenditures

4% depreciation and replacement of equipment

3% shareholders

10% Taxes

25% LABOR

50% SUPPLIERS (purchased parts, raw materials, and supplies)

Fig. 3-1. A simplistic view of corporate economics.

Table 3-1. Hypothetical Balance Sheet.

GOLDEN WIDGET CO.
Balance Sheet as of June 30, 1991

Assets

Current assets:		
Cash		$ 815,265
Accounts receivable		1,120,650
Inventories:		
Raw material	278,121	
Goods in process	204,164	
Finished goods	788,588	1,270,873
Prepaid assets:		77,873
Total current assets		3,284,751
Property, plant and equipment cost less allowance for depreciation of $935,482		1,324,835
Investment in subsidiaries		52,262
Deferred development expenses		78,984
Total assets		$4,740,802

Liabilities and Capital

Current liabilities:	
Accounts payable (trade)	342,600
Provision for pensions	162,400
Provision for income taxes	408,800
Other accrued liabilities	250,100
Total current liabilities	1,163,900
Provision for deferred income tax	92,405
Common stock	365,981
Paid in surplus	2,198,074
Retained earnings	920,442
Total liabilities and capital	$4,740,802

funds is also required to achieve a robust distribution activity. The balance sheet of the hypothetical Golden Widget Co. has been drawn up in TABLE 3-1.

Let us assume that you are an investor who would like to purchase a substantial interest in the company, in the way of shares, because it is a public concern and its shares are traded over the counter. Widget sales have been pretty strong and are running approximately $12 million on an annual basis, with profits after taxes of about $600,000. The balance sheet shown in TABLE 3-1 indicates that the company has total assets of over $4 million, which includes some $800,000 in cash.

The Golden Widget company has total assets in land, buildings, equipment, etc., of about $5 million. As an investor, it would be nice to be able to have your company make a profit of $600,000 with fewer assets, but after some questioning you find that if the company were to get along with less plant and equipment there would not be enough capacity to keep their customers supplied and happy. You notice that Receivables total over $1 million, but it has been the company's experience that pressing their customers for faster payment usually results in somewhat depressed sales, and the company needs the added volume to remain cost-competitive, simply because the widget market is extremely sensitive to price considerations.

In addition, you mention to the company's accountant that the "finished goods" inventory appears to be high, and might it not add to profitability to reduce this amount, since carrying the inventory costs money? She explains, rather patiently, that, because of the wide range of models and sizes, their particular business requires a sufficient stock on hand for most models, and besides they have a stock turn of about 6, which in their industry is exceptionally high. Altogether, the loss of customer goodwill would neutralize any cost savings that might be achieved by reducing the finished goods inventory.

Further questioning into the raw material and goods in process inventories indicates that with a fairly complex manufacturing process, goods in process of approximately $204,000 is needed in order to keep full employment and to stabilize the work force. The highly skilled labor force required to produce widgets has been developed by the company over the years, and workers with these skills are not replaced easily in this labor market. Also, because of the many variables in raw materials supply and with wide fluctuations in price, the raw materials inventory of almost $300,000 represents the efforts of the purchasing department to take advantage of quantity discounts when raw materials are offered at their lowest prices.

The results of your probing lead you to infer that the company, by and large, is using all of its assets productively. Therefore, you conclude that if any reductions were to be made in any of the areas of capital investment, the result of lower investment would probably be offset by higher costs, lower sales, or a combination of both.

A company's assets, in the main, can be divided into two categories: manufacturing and nonmanufacturing. Therefore, if the manufacturing assets have been pared to the bone or have been kept quite lean, then we must look to the nonmanufacturing assets to increase profitability.

In general, most companies realize that they are in two (at least two) different businesses: one is manufacturing and the other is marketing. Very often the marketing, or nonmanufacturing, aspect of the company is even more important economically than the manufacturing stage. Companies that are marketing-oriented, such as the cereal and cosmetics firms, typically have vast investments in advertising and marketing. From an organizational and tax vantage point it is even feasible for companies such as these to split into two different entities: one a manufacturing company and the other a marketing and distribution concern.

These two concepts have gained fairly widespread recognition over the years, but it is only just recently that a third economic activity is being carried out in their organizations, one that permeates the entire operation, namely the activity now known as materials management. This is true even in the service and utilities industries; for example, the Southern California Edison Corp. (SCEcorp) announced in its July 19, 1990, second quarter report that its board of directors had made three personnel changes, one of which was that Charles B. McCarthy, Jr. had been elected Senior Vice President and made responsible for Fuel and Material Management, Power Supply, and Engineering, Planning, and Research.

The activity of materials management, while less known or familiar, meets the value-added criteria discussed previously, since capital is employed in the purchase of materials handling and storage equipment, and costs are incurred to produce articles of value.

Adding value in materials management

The capital investment required to conduct an efficient materials management program is obscured and virtually hidden from the general view because, as sometimes happens with the fairly obvious, we can't see the woods for the trees. It is very much like standing an egg on its end—unless you crack the shell in order to make it stand upright, it never will. Thus, it is that the capital employed in materials management is buried in the materials inventory—that is to say, raw materials, semi-finished materials, purchased-finished materials, and the products that are produced and stored by the company. Added to this is the capital required to purchase materials-handling equipment; buildings to house the inventory; land for the buildings to stand on, if not leased; and office space and equipment used by the materials handling staff.

In general, since materials management, or phases of it, have since time immemorial been lumped together with manufacturing, the capital used for materials management has been buried within the manufacturing department and has not been permitted to stand on its own two feet. Thus it is that a poorly run materials management miscellany can be submerged within an effectively functioning manufacturing organization because manufacturing overshadows the activities carried on in materials management. Manufacturing operations transform the raw and other forms of materials that are brought to its various operations by the materials-handling operatives, who are a part of the materials management entity, and in this manner assist in the process of adding value to a product.

As an example of the hidden costs that can be reduced in materials management operations, there is the company that had press brakes, shears, and stamping presses in a large building. At one end of the building, which had railroad tracks entering it, flat plate or sheet steel and coiled steel were stored. The storage area was served by a couple of 50-ton cranes and hoists on overhead craneways. Steel was picked up off railroad flatcars and open gondola cars and piled in the storage bays, as it came off the cars. Sheet steel was separated in each stack by heat and grade, using wood dunnage to keep each load separated from the other. Coils of steel, some weighing as much as 20 tons each, were piled in pyramid fashion, as were loads of barstock of all different sizes and heat treatment.

It was very difficult, labor- and time-consuming to destack coils to obtain the right coil, or the right sheet steel for the presses and shears, whenever a particular size, thickness, or temper was required. The material control crew could be counted on to obtain steel that was not buried too deeply by resorting the stack; if too much moving and rearranging was required they notified the department foreman. As a result, the foreman in charge of the operation would call the local steel supply center and obtain a supply to keep his presses running. The steel supply house was located a few streets below the plant, and grew fat and happy as a result of the poorly organized steel storage facility.

A suggestion was advanced to install cantilevered barstock racks in the storage area, and to store coils in similar heavy-duty racks so they could be stored and removed individually; in addition, it was suggested that the sheet steel be placed in racks, also, to facilitate order picking. The return on investment for this installation was calculated to be slightly under two years for the racks, including the labor to effect the change and the additional materials-handling equipment required to handle barstock.

The impact of purchasing

In the previous example, which is a factual account of a real-life situation, if the accounting department, the purchasing department, the plant manager, and the material control supervisor had been paying attention to costs and analyzing the activities of this facility, the situation would never have been permitted to become the nightmare problem that existed when I happened across it in my first visit to this operation.

The first bell and whistle that should have been sounded, no doubt, should have come from the material control department. But, simultaneously should have come warnings from the accounting and purchasing departments. For the sake of expediency let us, for the moment, leave out the required actions from accounting and the plant manager, and go on to the role of purchasing under a capable materials management organization. The cost of purchased parts and materials from outside suppliers is the single greatest cost of the product (in almost all companies).

It stands to reason then, that purchasing is the single most important materials management function in almost all companies and institutions. Since this is so, then it is apparent that the terms *purchasing* and *materials management* have been used synonymously, despite the fact that there are more activities involved in materials management than purchasing alone. Managers who have purchasing as their sole responsibility are sometimes called Materials Managers, and other executives, who are known as Directors of Purchasing, are sometimes given additional responsibilities that make them materials managers despite their titles.

When the named and unnamed materials managers succumb to the impression that they are materials managers, it is unfortunate that in almost every instance they are not properly organized to perform that function satisfactorily. The fact of the matter is that purchasing is a specialized part of materials management, just as planning engineering is a specialized part of manufacturing and the preparation of trial balances is a specialized part of accounting. When the purchasing operation is completely synchronized with the other activities of the materials management function, cost savings will be made because inventories will be kept in balance; good buys will be made at relatively low prices; and the redundancies that were noted in the steel storage facility will be avoided.

The materials manager has the responsibility not only of obtaining materials from suppliers, as in the purchasing function, but also of marshaling these materials up to, and through, manufacturing operations. After the manufacturing processes have been completed, it is the function of materials management to see that the finished product is guided through the physical distribution channels of warehousing and transportation, etc., to the ultimate consumer. Thus, the function of materials management in obtaining (purchasing) and transporting materials from the supplier to the point of use (manufacturing), is reversed after manufacturing is completed, in the process comprising the physical distribution function. The same events take place in nonmanufacturing enterprises; however, instead of manufacturing you could substitute patient care in a hospital, or eleemosynary institutions.

How marketing is related to materials management

As the salesman knows, nothing happens until somebody sells something; therefore, the marketing department's task is related to all those activities that must be carried out in order to achieve the goal of selling the company's output. In some companies this might entail surveying potential customers to determine needs, for example: what size, horsepower, configuration, types of function, etc., are required of a product. Also, what is the population of the market for the company's product? How many units can be expected to be sold? How far into the future can sales be adequately forecast? What kind of advertising program is necessary? What colors are required? How should the product be distributed? Sales price? And, so on, through a gamut of interrelated subjects, with the principal effort being expended in getting the customer to buy the product.

Since the product must be stored, handled, and transported to achieve the company's marketing objectives, virtually the same physical processing must be performed that was a part of the purchasing function in bringing the raw, semifinished, or finished materials to the first point of use in the manufacturing cycle. This process of moving materials, or products, to the ultimate consumer goes under the heading of *physical distribution*.

The technical know-how associated with physical distribution is virtually the same as the skill required by the materials manager in obtaining materials and guiding, handling, and transporting them through the manufacturing process, with the sole difference being that while the materials manager sets policy for the earlier stages of manufacturing handling, the marketing manager establishes policy for physical distribution, which is carried out by materials management. It might be part of the marketing strategy to floor-plan, or consign product to the customer rather than store it in the warehouse, etc. The materials manager then has the responsibility for ensuring that the right quantity and type of product is transported and delivered to each customer.

Materials management and manufacturing

In some companies, it is very difficult to determine where materials management and manufacturing functions begin and end. The relationship between the two organizations and their respective functions is extremely complex. In essence it is more complex than the differences between marketing and materials management. In some instances, the relationship between materials management and production is clear and well defined; for example, in a newspaper publishing company, materials management will purchase the paper, ink, chemicals, etc., and deliver them to the presses. From that point on, production takes over and the resultant product—newspapers—are then placed into vans, trucks, and railroad cars for delivery to the ultimate consumer.

In most manufacturing, however, the relationship between manufacturing and materials management becomes complicated because materials can be purchased in various stages of manufacture. Materials can be purchased raw, semifinished, or finished and become part of a larger component or assembly. Then again, purchasing can buy a complete product, bypassing manufacturing completely, in the "make-or-buy" category of commerce.

Breaking down the make-or-buy decision into its component philosophies, whether the company makes a part or an entire product, or buys a part or the entire product, becomes a purchasing decision that usually requires approval by the company's top management echelons, depending upon the dollar amounts involved. Usually the dollar limits involving make-or-buy decisions have been established by management as guidelines for the purchasing department; however, in almost all make-or-buy events the materials management department operates in conjunction with manufacturing. In other words, whether to make or to buy requires the input of the manufacturing department, since this is not an arbitrary or unilateral decision of the materials management organization.

It is quite apparent, therefore, that materials management has a line function responsibility for any material that is purchased, up to the point where value is added by manufacturing. During the processing of the material, the materials management organization serves manufacturing in a staff capacity in almost the same manner it services marketing; i.e., storing and transporting, and providing materials-handling effort. Although the final responsibility for everything that happens during the manufacturing process rests with the manufacturing manager, since the materials management department is concerned with a great deal of the materials-handling activities there must be a large degree of cooperation and coordination between the two functions if the optimum cost-effectiveness of the overall plant operation is to be achieved.

4

Purchasing practices
in materials management

Introduction

The present status of materials management in industry and service organizations is relatively spotty. This deviation from the full-scale materials management concept is due in part to the way in which the practice has developed over the course of the years. Unless a company or service industry has just recently been formed and knowledgeable practitioners have installed the materials management concept in toto, it is most likely that a process of evolution has been employed within the company to the extent that bits and pieces of the materials management concept have been placed into effect, with the result that materials management can be considered to be in a state of development and has not fully matured into a total materials management philosophy.

Since the practice of materials management has evolved gradually over a period of time, it is only fair to point out how it has developed. Some form or other of materials management takes place when managers of a company take on, or acquire, additional functions. For example, a materials control manager might take on the duties of materials handling and warehousing, or vice versa; however, the additional duties are not his or her major concern. Other groupings of functions might occur in the company, such as the purchasing agent (in smaller companies) and the director of purchasing (in larger organizations) assuming responsibility for storage materials. As a result of these successive groupings within the company, the titular head—i.e., the president, or general manager—becomes for all intents and purposes, but largely by default, the materials manager, simply because all of the group heads report to him/her.

The problem with this arrangement is that the company president is only partially aware that he or she has become a materials manager. When this realization surfaces, then it is time to orient the company toward the full-scale organization that is required for a cost-effective materials management program.

The company president usually relies very heavily upon his or her director of purchasing to implement the materials management program, although some concerns have gone outside the company for assistance in installing materials management programs, either by hiring a materials manager or by engaging a consultant. In chapter 1, FIGS. 1-2 through 1-5 illustrate typical organizational structures that show how the materials management concept has developed from the traditional organization concept (FIG. 1-1) of a manufacturing company. Since the largest number of materials management applications have occurred in manufacturing, parallel developments in the service industries and the military establishment have not been represented in this text, but they have taken place, nevertheless.

In many instances, the director of purchasing has been "kicked upstairs," so to speak, and has assumed the title of materials manager, as well as most of the functions and responsibilities of the position. The head of purchasing's task, in any manufacturing company, is a prestigious position with considerable clout, since, usually, half of the company's expendable funds is involved in product cost. In addition, the purchasing department must maintain close coordination with manufacturing, marketing, and physical distribution. It is this very close association, required by a well-organized and successful company, that permits us to borrow a definition from the military for the term *logistics management* and call *materials management* the science of procuring, maintaining, and transporting materiel (and personnel in the case of the military).

The first real materials management philosophy from which today's concept is derived was practiced by generals who launched large-scale warfare—from the Egyptians, the Hittites, and Alexander the Great to Napolean, our own General Washington, and down to present-day warriors. Not unsurprising is the fact that, when supply lines broke down or when procurement was slow in responding to needs, all manner of problems and reverses were experienced by the military. Examples of these debacles are the lengthening supply lines that formed Napoleon's retreat from Moscow and George Washington's sad winter at Valley Forge. To quote from Napoleon, "An army travels on its stomach," and the story of successful military campaigns are triumphs of successfully carried out (logistics) materials management programs. Outstanding examples of carefully planned and brilliantly executed materials management programs have been the Allied Forces' invasion of France during World War II, starting with the landings on the beaches of Normandy, and the most recent war on the plains of Iraq.

The evolution of purchasing and inventory control

From cottage industry to business administration is a quantum leap, simply because the tools of business have changed extraordinarily. It is like comparing a Ouija board with a mainframe computer. In the dim past, say at the turn of the century, the plant superintendent or the shop foreman—the "boss"—scheduled production, did his own purchasing, controlled inventory (usually by eye-balling his stock), and did his own hiring and firing. In those days, it would have been considered ridiculous to have suggested that the plant could do better with an individual who was employed solely for the purchasing function, or inventory control. Nevertheless, as the business expanded and as purchasing became more complex since engineering specifications began to enter the picture, it was necessary to segregate this function from all the other activities of the plant manger.

This emerging function of purchasing as well as inventory control, or managing inventory, surfaced primarily as a result of the recognition that different skills are required to purchase than to specify materials and to control inventory. Engineering skills are needed to specify materials, but buying requires some legal training and an understanding of the capabilities of vast numbers of different companies. Inventory control developed a sophisticated formulary based on statistics and the mathematics that promoted linear programming and regression analysis, and thus became a separate, but integral, part of materials management.

The professional purchasing agent can almost always procure materials at a better price than most nonprofessionals; therefore, since purchased materials comprise such a large part of product cost and company expenditure, the trend in business management has been to give the purchaser increasing status in the organization. This visibility becomes of value to the purchasing department since liaison with finance, engineering, manufacturing, and marketing is especially necessary in order to carry out the objectives of the department and to promote the concept of *least total cost*. The closer that purchasing and manufacturing work together, the fewer the problems that might arise concerning the quality of parts and delivery schedules—two all-important considerations for manufacturing that have a direct impact on company expectations.

Liaison with engineering is considered a vital part of the purchasing department's activities since many companies often involve the cooperation and assistance of suppliers in helping them develop new products and parts. Again, certain suppliers develop a tremendous amount of expertise in the fabrication of certain parts. As an example, in the automotive industry, even engine manufacturers have suppliers whose special capabilities lie in the manufacture of valves (stellite valves of particular durability) and aluminum piston heads, whose uppermost cavities require exacting and tremendously precise machining capabilities.

Many of the manufacturing practices of companies such as these are trade secrets that have been developed over a period of years at tremendous expense. For instance, there are only four companies worldwide who manufacture the bulk of piston heads for all the internal combustion engines used in passenger and industrial vehicles; similarly, there are companies that make ceramic and plastic parts with a great degree of precision, in addition to many other companies that fabricate only parts and components and never an end product.

Purchasing has gained stature and position in the corporate world. No longer is it looked upon as a necessary evil in the scheme of things, but as an entity that has a great deal of impact on corporate success and the margin of profit. This is true not only in manufacturing but also in nonmanufacturing organizations, such as transportation, public utilities, medical care facilities, and governmental activities on the state level.

For these reasons, the purchasing department must be kept informed regarding company policy, marketing plans, and engineering developments. The very fact that purchasing has such an important impact on various crucial areas of the company's operation makes yet another reason why the head of purchasing should be a member of the company's top executive group. When the top purchasing individual reports directly to the president of the company, the materials management organization directly under his supervision (or whose materials manager reports directly to him or her) becomes all the more effective because of the clout it has acquired in the organizational structure.

In many companies, because of the recognition given the purchasing function, representatives from the purchasing department are included in every group concerned with scheduling, inventory or production control, production planning, cost reduction, standardization, metrication, new products, and make-or-buy decisions, to name a few. In fact, a well-organized and effective purchasing department with a good top man or woman, will make all major buying decisions within the company. A good purchasing department will also introduce members of the engineering department to suppliers with special expertise or to new products that will enhance the productivity and effectiveness of the engineering staff. It is not uncommon for members of the purchasing department, whose contacts are legion, to make recommendations or suggest design changes to the engineering department.

One of the principle areas in which the purchasing department has considerable effectiveness lies in the area of design specifications. Many engineers will express tolerances with an unnecessary degree of precision, for example, a shaft with a diameter measured in .0001± inches, when .001± would be sufficient. Many an engineer, me included, has been guilty of requiring a degree of precision beyond the functional scope of the part and has been saved from embarrassment by a knowledgeable veteran of the purchasing department who has procured the same or similar parts for years and understands the use and function of similar parts. The higher degree of (tolerance) precision can spell the difference between a part costing dollars more versus one amounting to hundreds of dollars more in cost.

For reasons such as these, in many companies the purchasing department will make all major buying decisions and certainly have a significant impact on questions of make-or-buy. Another area in which the purchasing department often proves invaluable is in suggesting engineering design changes to enhance the profitability of the product, since if components can be changed somewhat, if only slightly, they can be purchased at less cost.

Also, combinations of parts, or component subassemblies, sometimes can be subcontracted to other companies with special expertise or slack time, who will manufacture or assemble parts for the company merely to fill in during lulls in their own production schedules, mainly to avoid layoffs or to build up their production volume. Sometimes, the purchasing department will bring in new vendors or will introduce the engineering staff designers to new products so that they will not have to redesign the wheel, so to speak. Some suppliers will do a certain amount of free engineering just so that they can either retain an established relationship with the company or, if they are a new vendor, can introduce themselves and their capabilities with the hope that they will be able to sell their wares to the potential customer.

Purchasing must also work closely with the inventory control department because buying in economic lot sizes is part of the purchasing department's stock in trade. The inventory control phrase, *E.O.Q.* (economic order quantity), is still a much debated and formulated subject, especially since the Japanese have publicized Just-in-Time manufacturing, where the best inventory control is tantamount to no inventory at all. Today's effective purchasing department makes use of all the devices imaginable in assisting the inventory control department in achieving its objectives. For instance, special arrangements will sometimes be made with suppliers who have been qualified and whose reliability is unchallenged and proven, in order to maintain safety stocks of parts and components in their own plants, which are shipped to the company at prescribed times, on a fixed schedule, or on an emergency basis, depending on the authorization of the purchasing department.

The basic elements of purchasing

While the purchasing department will maintain close lines of communication with engineering, marketing, and finance, there is an even closer rapport with inventory and production control, as well as with the traffic department. In almost every company, the combination of these three departments effectively performs most of the materials movement that takes place at the company's behest. An enlightened top management will almost always subscribe to the concept of materials management under the direction of one manager, since this choice focuses direction and establishes responsibilities, rather than diluting them.

For the effective, first-rate purchasing department, the liaison aspect of its work will be an important, but necessarily subsidiary part of its charter. The primary function of the purchasing department is to buy materials usually in the amounts requested by production and inventory control, and the warehousing entity.

The activities of the purchasing department can be grouped into roughly the following categories:

- *Buying*—Finding and qualifying suppliers, negotiating purchasing contracts, negotiating terms of purchase that will be mutually advantageous to both the company and the vendor, issuing purchase orders, and buying against contracts when the vendors default.
- *Delivery*—Expediting delivery from suppliers when due-ins are slow in arriving, or when necessary to ensure delivery in time to meet critical manufacturing schedules; negotiating any new terms and changes in purchasing schedules when necessary to fit changes in manufacturing schedules, or when circumstances or business conditions vary.
- *Liaison*—Serving as liaison between suppliers and any of the company's departments, such as finance, marketing, engineering, manufacturing, production and inventory control, quality control, and any management group including accounting and computer operations.
- *Information*—Serving as a place where manufacturing and engineering can obtain information on new products, new processes, types of suppliers, and changes and events in business and market conditions that can affect the sale of the company's product.

It is important to note, at this point, that the ubiquitous computer has both aided and made more complex the intricacies of the purchasing function as we know it today. The reason for this apparent paradox is that it has broadened the scope of the purchasing function in its relation to materials management. Thus, by providing more powerful tools to work with it has required more conscious planning effort, and as a result, the task of purchasing has been placed on a more scientific level.

Modern-day purchasing departments are a sea of computer terminals, and each buyer in the department has a wealth of historical information and data to work with. It is possible to have not only lists of suppliers, but the latest prices, best quantity discounts, prices, market information, supplier performance, and the like at one's fingertips.

In addition to the buying information, the computer data bank makes it possible for inventory position to be learned on a real-time basis. Of course, printouts of reorder quantities, the buy notice, safety stock, lead times, and the like can be obtained on a monthly, weekly, or daily basis depending on the characteristics of the firm's business.

The proliferation of computer technology has had an enormous impact, not only on the way the purchasing department conducts its business, but also on the organization of the purchasing department, especially in its relation to the materials management concept.

Purchasing organization

In many companies, the materials function, materials management, or the movement of materials is under the authority of, or subordinated to, manufacturing. In other companies, however, materials management has been separated from manufacturing because top management feels that it can make a major contribution to profit if it is an independent entity—not so much an independent function, since its existence depends to a large extent on how well it satisfies and anticipates the needs and requirements of manufacturing. Severing materials management from manufacturing enables the manufacturing manager to focus more of his or her attention on automation and other technological concerns, thereby increasing manufacturing productivity.

The computer helps to integrate the functions of both manufacturing and materials management. Organizational charts help to delineate the distinctions between the two groups, and by examining these distinctions carefully, you can see that the materials manager has responsibilities that encompass inventories, materials handling, purchasing, warehousing, traffic, receiving, shipping, and, in general all materials movement, including production control and scheduling—or combinations of these things that have to do mainly with materials. Some of these functional responsibilities, in some companies, still remain under the aegis of the manufacturing department, but the trend is rather to lump all of material movement and materials under the heading of materials management and let the manufacturing department handle what it does best—i.e., manufacturing—and let the materials manager be concerned about the so-called peripherals of the operation.

Another impetus in this direction is the emphasis on automation and data processing in the factory, as well as in nonmanufacturing enterprises, which has propelled many companies to combine all of their materials management functions under a single manager.

If you were to break down the materials management task into its major components, a good case could be made for the following functions:

- Traffic
- Materials control
- Purchasing

In perspective, however, the overall materials job can be divided into a number of different functions, such as materials handling and inventory control, that can be grouped under the three major categories cited. The ideal organizational structure varies from company to company.

After an examination of several hundred organizational patterns in as many companies, I have come to the conclusion that the productivity of an operation is not dependent so much on the organizational structure of a company, but rather on the people employed by the company. Morale, esprit de corps, a work ethic, etc., mean more to a successful operation than a glamorous organizational chart, but in many organizations it is the starting place.

Since different managers and top executives view their organizational domains from differing perspectives, it would appear that the most feasible method for creating a materials organization is by the division of work, for example:

- Function
- Location
- Product
- Manufacturing process

A good argument can be made for any one of these four organizational divisions, and it would make good sense to examine each of them circumspectly to see how well they fit into your plans for organizing for materials management or how well they coincide with your experiences.

Of all the most widely employed devices, the organization based on function is the most accepted, simply because tasks that are organized on function promote the maximum use of specialization and job skills. In a functionally organized department, tasks are diluted to the extent that each job becomes very highly specialized, and skills are concentrated into a very narrow spectrum. Each worker becomes exceptionally expert in a very specialized area. Importantly, there is a considerable degree of functional specialization in every materials organization; thus, separating the overall materials activity into its component parts, we have materials handling, purchasing, traffic, inventory control, and so forth, all of which are forms of functional specialization.

Furthering the analogy, within each of these groups, there is even more specialization. As an example, purchasing can be subdivided so that each person within the department has specialized tasks such as receiving purchase requisitions, contacting suppliers, printing out purchase orders, and the like. Each specialist handles a portion of the overall job so that in a large purchasing department, the buying job itself might be composed of dozens of people: different buying specialists (buying a variety of commodities), assistant buyers, expediters who chase due-ins, purchase analysts, and others.

When considering a materials organization, it is realistic to divide all of the work entirely by function. This method would work extremely well in most small organizations; however, there might be some difficulties in a larger operation, although the problems that would arise would probably be minor.

In establishing function as a basis for a materials management organization, job skills can be divided so that higher-skilled jobs are not diluted. For example, the purchasing agent who heads up the group would have a number of buyers and assistant buyers. The buyers as a class would be serviced by a clerical staff to perform the purely clerical operations, such as filing and typing.

Functional specialization would make it economically feasible to have several departments use the services of the company's economist (in larger companies); that is to say that marketing and purchasing could share in the economic forecasts provided by the economist. If this is not possible, then the economic forecasting performed by the marketing department can be made available to purchasing. Wherever economic forecasting is accomplished within the company, the results of this labor, because it is highly specialized, can be shared by all departments who require this type of information.

Employees in a department where work is allocated on the basis of functional specialization become extremely efficient at each assigned task where repetition is a factor. There are several advantages in this high degree of specialization. As the business expands it is easier to recruit and train employees in jobs that are highly structured. In times of business contraction, employees with the lower skills can be transferred out of these departments into other areas of the company where the learning curve for new jobs is of relatively short duration. Thus, organizing a department by functional specialization has certain advantages over other forms of organization.

Another advantage is that in any discussion of employee satisfaction and dissatisfaction, it has become rather conventional to insist that highly repetitive tasks increase job boredom. Although this might be true of some overqualified achievers, the same cannot be said with any degree of certainty when applied to clerical tasks in a purchasing department, as an example. It has been noted that employees who fully understand their tasks take great satisfaction in accomplishing even the most routine and repetitive tasks well. A parallel can be established between this type of job and driving a metropolitan transit bus. An applicant with less than a high-school education is more favored than one with a high-school diploma. Applicants with more than a high-school education are discouraged from applying.

In the organizational chart of FIG. 4-1, the overall management task is shown as it is subdivided into its major functions. In this organizational structure, the third level of the chart depicts the functional elements of materials management, which have been separated into materials control, purchasing, and traffic. It is possible, with this form of organization to accomplish the type of functional organization within each department that has been accomplished with the plant, as a whole. This type of functional division works well for many companies, even large multinational corporations.

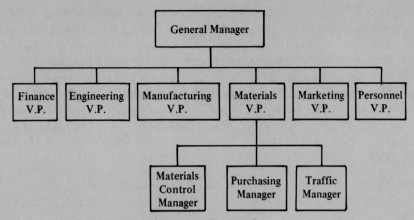

Fig. 4-1. Organizing a manufacturing plant by function.

Some multiplant companies, on the other hand, have seen the need to use the "location" factor in establishing counterpart organizations in each plant, especially when business expansion dictates the location of other plants usually in geographically dispersed areas. Sometimes the reason for doing so is that demographics will determine centers of population that make physical distribution a viable factor in plant location. The notion is, let's take the manufacturing capability, or whatever the product or service is, to the customer instead of operating at a remote location and having large transportation costs nibbling away at profit.

Other reasons for locating plants at geographically dispersed areas are more devious; for example, labor-intensive industries disperse the manufacturing capabilities so that all the eggs are not in one basket; in other words, separate the labor force into different labor jurisdictions and/or labor contract periods so that no one labor contract can affect work stoppages simultaneously. With today's mass communication and media, this type of corporate strategy is becoming less and less effective for many industries.

Some elements of the multiplant company's organization must be replicated at the local plant level even though the principal authority for the department remains at the corporate level in the headquarters building, or plant, of the company. Figure 4-2 illustrates a division of effort by plant location.

The corporate materials manager's chief function is to assist the plant materials management organization by establishing operating policies and, in general, evaluating and directing performance. The performance in question is of the plant materials manager and not of his or her materials management organization. Usually, salary and perquisites are discussed with and mutually resolved among the plant manager, his or her personnel manager, and their corporate equivalents in the main office.

Salary and wage administration has developed into a highly skilled and high-powered subject in most companies. Many companies hire consultants who have established reputations in this area, some of which are deserved. Many other companies have become mired in the complexity of formulas that sound convincing on paper but leave much to be desired in practice. Much company strife has developed as a result of an arcane science poorly practiced.

By and large, there are many other areas (departments) that are replicated in each plant, for example, purchasing. If we substitute purchasing for materials management in FIG. 4-2, we find a similarity based on location; however, in many companies this duplication of effort sometimes gets out of hand. This happens when the entire department is purchasing the same or similar items that are being purchased in the main plant or the corporate office. What happens when this occurs is that the vendors or suppliers often find that the plants are bidding against each other for similar items. When plant size is small, this is of no great consequence, but when individual plant requirements become large, as in some multinational concerns, the marketplace becomes an auction block.

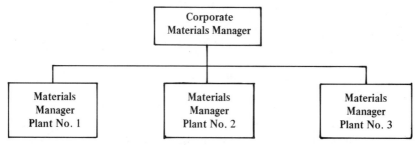

Fig. 4-2. Replication of corporate functions by plant location.

In order to avoid the excessive costs of decentralized purchasing wherever several plants of the same company are buying similar materials, it is much more cost-effective to combine the purchasing into one centralized division. This is not to say that each plant must depend completely on a centralized authority for all its purchases; however, where massive buying is performed for specific commodities, large cost savings can be made and improved delivery schedules can be maintained for these items if the buying is centrally located. Smaller purchasing requirements, which might be of particular usefulness to the particular plant, should be made at the plant level.

Although the purchasing activity for a materials department can be organized according to "function" or "location," a third method is to organize the department by product, process, or project. As you can see in FIG. 4-3, each buyer within the company is responsible for a given product; however, "process" or "project" can be substituted for "product."

From an organizational standpoint, there is very little difference between a purchasing department that is set up around product lines and one set up around manufacturing processes. These are relatively permanent organizations that will probably endure as long as the company remains in business.

The project type of purchasing organization, however, is relatively short-lived. Project organizations are usually established on a crash basis to carry out a particular task. Thus, when the task is completed the members of the group return to their original organizations or are recombined into a new department.

As an example of a project-type organization, let us suppose that the company wants to introduce a new product or product line. It would select a project leader, and

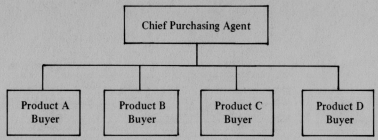

Fig. 4-3. A purchasing department that is organized by product.

then, with his or her assistance or with a project committee's guidance, potential members of the group would be selected from manufacturing, engineering, materials management, and so forth. In many instances, after the product has been introduced the members of the group would either return to their former jobs or be assimilated into the new organization to assist in the start-up and running of the new production division.

In many materials management departments, the distribution of work in the purchasing organization is often accomplished on the basis of product or commodity. Each buyer within the organization might have a certain grouping of commodities for which he or she has sole responsibility. In this regard, it is really a combination of functional specialization and product specialization. For example, one buyer might buy all the paper goods; another, rubber products; and another, all the electric components.

There is a very real advantage in this form of specialization because the buyer becomes relatively expert in his area of purchasing. Buyers often visit manufacturers' plants, and in discussing purchasing specifications with the suppliers' personnel, the expertise the buyers acquire is invaluable to the purchasing company. Another advantage in having the buyer become more closely acquainted with his or her product, or commodity, is that he or she will be better able to relate to the profitability objectives of the company. He or she will do a better purchasing job, and his or her morale will be enhanced by the ability to travel at company expense to the vendor plants.

In some larger companies, the purchasing organization might be subdivided into buying groups that purchase materials for each division of the company. For example, a buying group would be devoted to purchasing foundry materials for the foundry division; another group would buy sheet and coil steel for the stamping division. In this type of organization, the division of buying groups into process or stage of manufacture would require that each buyer be responsible for determining requirements, managing inventories, printing out purchase orders, and doing his or her own expediting to ensure that stock levels are maintained at some established point.

In smaller organizations, which might be responsible for only a single product, a small product line, or a large number of small product runs, such as in job-lot manufacturing, the organization by product is not too logical a division of responsibility, and the *process*, or stage of manufacture, offers a great deal of advantage for purchasing effectiveness. Figure 4-4 illustrates a purchasing organization showing buying divisions by process.

Despite the different types of organizational structures that have been developed for the purchasing function, many of the larger companies still rotate buyers from one

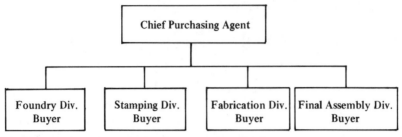

Fig. 4-4. Organization of purchasing by processing division.

commodity to another after a tour of duty comprising two to three years. One of the principal reasons for this transference is that the buyer and vendor relationship might become too close for real objectivity on the part of the buyer regarding the vendor and his product. Another reason is that having the buyer acquire additional expertise with new commodities increases the flexibility within the buying division. A buyer transfers the tools of his trade to the new area and in relatively short time he is quite capable in the new commodity group.

Some buyers, however, have greater skills in some fields than others; for example, a steel buyer becomes more at home with metallurgy than he would be with textiles. In large companies, however, where rotation of buyers is practiced, rarely would the transfer be this extreme.

In the early stages of a company's growth there might be only one buyer, who is usually given the title of Purchasing Agent. As the company grows, the P.A., assuming that he or she is competent in his or her job, hires one assistant, called a Buyer. The business continues to grow and several more buyers are hired. Eventually, the purchasing agent has so many buyers reporting to him or her that his or her *span of control*, to use a military term, has exceeded his or her ability to manage, or at least to keep in communication with all of the buyers. In order to effect an orderly transition to a more manageable entity, our purchasing agent reorganizes his or her department with the blessings of top management. The P.A.'s title might be changed to Director of Purchasing or he or she could simply remain as P.A., but he or she is the head of the organization and has successfully shortened his or her span of control, as you can see in FIG. 4-5.

Although the organization chart of FIG. 4-5 indicates that there are 34 slots or positions in this organization, there might be at least twice that number of people because of the necessity to employ a certain number of purely clerical assistants, such as file clerks, secretaries, typists, and word processors. In the future, as office procedures become more streamlined and technically advanced, it will be entirely possible to eliminate much of the paperwork that is used to communicate with suppliers, engineers manufacturers, etc. The technology is here to do so; it is solely a matter of putting it into practice and installing the hardware and the software (computer equipment and peripherals) to do the job.

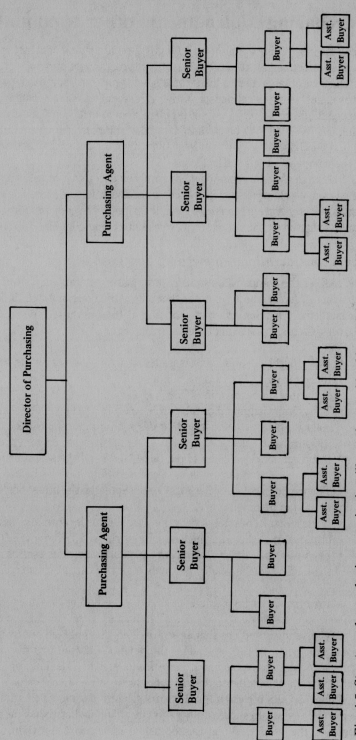

Fig. 4-5. Structure of a purchasing organization, illustrating considerable depth but a relatively narrow span of control.

Purchasing contracts and other legal matters

Since the purchasing agent or buyer signs the purchasing contract or agreement, he or she is in effect, an agent of the company. Although most purchase orders and contracts contain a lot of language that is termed *boiler plate*—i.e., standard legal paragraphs that cover almost every legal aspect of the agreement—in some instances special circumstances might arise. Wherever a problem arises, or other than a routine purchase is being made, it should be reviewed by the company's lawyers. In many instances, printed purchase order forms have standard legal clauses that should also be reviewed by the company's legal staff.

Inasmuch as the written (printed) purchase order will, in every instance, take precedence over verbal agreements in a court of law, it is sometimes necessary to have the results of every oral agreement reduced to writing. It is good practice, also, to have those results reviewed by the company lawyer before being committed to the agreement.

Legal considerations concerning purchase orders

The bulk of the company's purchase orders are generally prepared in the purchasing department. Every purchase order that leaves the company and finds its way to a vendor is a legally binding contract. Looking at the purchase order process, we find the following events, causes, and effects:

- Every purchase order that is issued is an acceptance of the vendor's offer.
- If the terms and conditions contained in the purchase order vary from the vendor's quotation, then, in effect, it is a counteroffer.
- The acknowledgment of the purchase order by the vendor seals the contract and makes it binding, indicating that the vendor agrees to the terms and condition of the purchase order.
- If the vendor rejects any of the terms and conditions contained in the P.O., then he or she is, in effect, making a counteroffer.
- The vendor's counteroffer must be accepted by the buyer before the contract can be said to be binding.
- If the vendor does not accept the purchase order formally, but delivers the merchandise that the order includes, he or she is accepting the P.O.
- Each purchase order change in terms and conditions that has been accepted by the buyer and the vendor becomes a part of the purchase contract.

Many manufacturing companies do not even bother to place terms and conditions on their purchase orders. They have established good relations with their suppliers, and any misunderstandings that arise are often straightened out by a simple process of negotiation between the buyer and the vendor. Lawsuits or disputes over purchase terms with conditions are much more frequent when dealing with commercial businesses than on the industrial, or manufacturing, side of contracts.

What to do when negotiations break down   Sometimes there does not seem to be any way to successfully negotiate a dispute between your purchasing department and the vendor. When this occasion arises, that is, when negotiations fail to resolve the problem, then there might be only one last resort to fall back upon. This measure, as drastic as it might sound, is to call in an impartial arbitrator.

The American Arbitration Association has a list of arbitrators that can be used to referee a dispute. Each of the two parties concerned selects someone to represent them, and the official arbitrator is a third party in the dispute or the disputants simply agree to bide by the ruling of the arbitrator alone. The arbitrator is usually an attorney, but sometimes might be an unbiased expert with a great deal of experience in resolving similar issues.

The arbitration consists of both sides presenting their testimonies to the arbitrator(s). The arbitration hearing is faster and less expensive than a trial in a court of law; and since the hearing is not a public affair, the internal affairs, trade secrets, and the like, do not become a matter of public record. Another advantage of arbitration is that complex commercial and manufacturing problems may be dealt with and resolved much more equitably than if they were aired in a lawsuit before a technically unqualified judge and an equally unqualified jury. In the latter case, it would be extremely difficult to achieve a really objective evaluation of the problem.

In the past, the ruling of the arbitrator was not considered binding, and either party was still free to sue the other if they were not satisfied with the results of the arbitration. Many states, however, have passed laws that make the arbitrators' ruling legally binding once two parties agree to submit their dispute into the hands of an arbitrator or an arbitration group, which they have jointly agreed to.

Since the occasion for disputes to arise is not as uncommon as it might appear, and since the effects of problems that arise could prove very costly, many companies include arbitration clauses in their purchase contract. The American Arbitration Association recommends that the following arbitration clause be inserted in purchase contracts. (The wording has been inserted in this text to give you some idea of its substance; however, before you use the textual matter, it is recommended that you obtain legal counsel in order to determine its appropriateness for your company.)

> Any controversy or claim arising out of or related to this contract, or the breach thereof, shall be settled by arbitration in accordance with the rules of the American Arbitration Association, and the judgment upon the award rendered by the arbitrator(s) may be entered in any court having jurisdiction thereof.

Governmental disputes Purchasing by the federal government, as well as local and state political bodies, is more often than not conducted in a goldfish bowl. The federal government has a vast body of legalistic regulations governing purchasing, filled with forms and jargon that is usually very difficult for the layperson to comprehend. It takes some time to become familiar with even the barest essentials of the requirements, and it is a paperwork jungle that scares most small business concerns from responding to most bids.

All you have to do is examine a federal procurement document to verify the truth of this statement. Reams of paper in procurement documents spell out, in vast detail, wage and hour restrictions and minority and women's business requirements. There are even pages of federal and military specifications (in the instance of Department of Defense purchasing) that may or may not apply to the matter in hand. Also, buried in this mass of procurement regulation, on perhaps only a page or two, usually at the tail end of the document, are the bid requirements. Usually they are written so poorly, often ungrammatically, that the supplier has to wonder whether the request for proposal is worth responding to.

Be that as it may, however, contract disputes involving the federal government also can be arbitrated if both parties agree upon this avenue of settling the dispute.

Company agent vs. supplier Most companies are not as rigid as federal, state, and local governments are in their purchasing regulations. Generally speaking, the larger the company, the less flexibility; nevertheless, it doesn't pay to disregard the legal aspects of purchasing responsibility in any event. The terms and conditions of the purchase contract should be defined as clearly and precisely as possible, not only to avoid possible future disputes, but to absolve the company from any possible liability that might arise.

The company's purchasing agent has the power to bind the company to a contract with a supplier. However, if he or she makes a judgment error—for example, overpurchases a quantity of materials that exceeds the company's ability to use in a reasonable period of time—it is not the purchasing agent who is responsible for paying the vendor for the material; it is the company, since the P.A. is an agent of the company, hence the term *law of agency*. Thus, the company is liable for the bill, not the agent, because it is assumed that the vendor acted in good faith in supplying the material. Therefore, if the company refused to pay for the material in accordance with the purchase order, or contract, the supplier would have legal recourse against the company. Most companies try to minimize the amount of damage that can be done by ineptness or error on the part of their buying staff by subjecting most of their purchases to a schedule of reviews. For example an assistant, or junior, buyer might have the authority to issue purchase orders up to $500 without review. Any purchase order amount from $501 to $1000 would have to be approved by a buyer. An amount between $1001 and $5000 would be reviewed by a senior buyer, and so on until ultimately even the president of the company would have to approve any substantial amount for which the company might be held liable.

Of course, the matter of responsibility for purchase material is, in reality, a two-way street. The staff member of the purchasing department cannot assume that the personal guarantee of the salesman does, in fact, mean very much. A salesperson who calls on a company and promises a certain price or a specific delivery time cannot be held accountable for the supplying company. The only time that a buyer can receive a binding commitment is when the contract that he or she has negotiated with a salesperson is confirmed, in writing, with the company that the salesperson represents.

Also, a number of companies do not employ salespeople, as such, but they are represented by a number of manufacturer's representatives in various geographical areas of the country. Some companies do not even try to geographically disperse this representation—for instance, there might be a number of salespersons (each an individual manufacturer's representative) in a given city, town, or state, depending on the product.

The larger companies who use reps tend to restrict their number in any given geographic area. They feel that by maintaining a certain degree of exclusivity they will be better represented. This does not always bear out the statistics, however, because in the final analysis it is the product quality and/or demand for the product that determine how well the item or product line will sell in any given area.

Companies that buy and sell locally—i.e., within their state—are regulated by state statutes. There is some small variation within the fifty states, but by and large the Uniform Sales Act that was followed and largely superseded by the Uniform Commercial Code (U.C.C.), has attempted to restrict any wide deviation in the regulations among the states. Nevertheless, national companies must be familiar with each state's commercial codes, despite the so-called uniformity that is the goal of the U.C.C.

As examples of the wide disparity that exists between purchase orders (as contracts) and proposals (as contracts), two companies' actual provisions are reproduced herein, in toto. The first company is located in the Pacific Northwest and has engaged the second company, located in Florida, to install a prefabricated building in one of its plants.

The purchaser, which we shall call, for the sake of simplicity, the "XYZ Company," has attempted to cover all of its requirements, liabilities, and so forth, in FIG. 4-6 (on pgs. 52 through 54). The obvious details, such as price, method of payment, etc., are not as obvious as they first appear to be. The supplier would do well to make a checklist of the requirements so that the wealth of detail does not become buried in the clerical or administrative measures that must be taken to comply with the purchase contract. Such a checklist would be, as follows:

Purchase contract checklist

- XYZ Company address.
- Purchase contract number.
- Purchase contract change number.
- What the Change number affects; i.e., price, added item(s), added notes.
- Payment remit address.
- Date of purchase contract.
- Buyer's name.
- Buyer's phone number.
- Buyer's mail stop number (in a large company).
- Sales tax status (what sales taxes apply).
- The contract is limited to the provisions of the contract without any additions, deletions, or other modifications that are not spelled out in the purchase contract.
- Delivery of goods (or provided services) is deemed conclusive evidence of the supplier's acceptance of the contract.
- Shipping point to which the merchandise is to be delivered.
- F.O.B. point.
- Payment terms.
- Fixed price contract.
- If there is an acknowledgment copy attached to the purchase contract, it must be returned to the *buyer* (noted previously) within fifteen (15) days of the date of the purchase contract.
- Any federal, state, local, or excise taxes must be identified as a separate item on each invoice submitted by the supplier according to this purchase contract.
- Freight, postage, or service charges must also, as in the matter of taxes, be identified separately.

✔The supplier must send one invoice marked "original," and one copy of the same to the XYZ Company address, Attn. Accounts Payable.

✔The purchase contract number, change number, and applicable item number (including part number and serial number) must be shown on each invoice, packing list, bill of lading, and all correspondence that pertains to the purchase contract.

✔Payments to be made, net 30 days, after receipt of a correct invoice (unless otherwise agreed to in the terms and conditions of the purchase contract and/or purchase contract change).

In addition to this checklist, the purchase contract should include the following:

- A detailed specification of each item required.
- The item number.
- The quantity of each item.
- A shipping schedule, showing the date the item is required and the quantity.
- The unit price of the item.
- The extended price for the item. If there is one, it is extended as one; if there are several, then the multiplier becomes the number of items to be shipped.
- The net purchase contract value.

PURCHASE CONTRACT   PAGE: 1 OF 5

PURCHASE CONTRACT NO: B 233234   DATE: 10/02/87

PURCHASE CONTRACT CHANGE NO: *1*   BUYER NAME:
CHANGE AFFECTS: *PRICE*   BUYER PHONE: MAIL STOP:
ADDED ITEM(S)   ADDED NOTES

SUPPLIER NAME:

                                                    THIS CONTRACT IS BUYERS OFFER TO SUPPLIER
                                                    AND ACCEPTANCE IS LIMITED TO ITS PROVISIONS
                        PAYMENT REMIT ADDRESS   WITHOUT ADDITION, DELETION, OR OTHER
                     ENDURE PRODUCTS, INC.   MODIFICATION.
                     7500 N.W. 72ND AVE.
                     MIAMI FL 33166   DELIVERY OF ANY GOODS OR PROVIDED SERVICES
                                                    SHALL BE CONCLUSIVE EVIDENCE OF SUPPLIER
SALES TAX STATUS: *WA SALES TAX*   ACCEPTANCE OF THIS CONTRACT.

SHIP TO..:

FOB POINT: *MIAMI, FL*

PAYMENT TERMS: *NET 30*
PURCHASE CONTRACT SUBJECT TO TERMS AND
CONDITIONS: *FIXED PRICE CONTRACT DO 1000 0070 (REV. 6/86)* INCORPORATED HEREIN BY THIS REFERENCE.
•••
SUPPLIERS ADMINISTRATIVE REQUIREMENTS:

IF ACKNOWLEDGEMENT COPY IS ATTACHED,   ANY FEDERAL, STATE, LOCAL, OR EXCISE TAXES   THE PURCHASE CONTRACT NUMBER, CHANGE NUMBER
IT MUST BE SIGNED AND RETURNED TO THE FOR ARTICLES OR SERVICES OBTAINED BY THIS   AND THE APPLICABLE ITEM NUMBER, INCLUDING
BUYER (IDENTIFIED ABOVE) WITHIN FIFTEEN PURCHASE CONTRACT MUST BE IDENTIFIED AS A   PART NUMBER AND SERIAL NUMBER MUST APPEAR
(15) DAYS.   SEPARATE ITEM ON EACH INVOICE SUBMITTED.   ON EACH INVOICE, PACKING LIST, BILL OF
                                       FREIGHT, POSTAGE, OR SERVICE CHARGES MUST   LADING, AND ALL CORRESPONDENCE PERTAINING
                                       ALSO BE IDENTIFIED IN THE SAME MANNER.   TO THIS PURCHASE CONTRACT.
                                       SEND ONE INVOICE MARKED 'ORIGINAL' AND ONE
                                       INVOICE COPY TO THE ADDRESS SHOWN ABOVE,   ALL PAYMENTS SHALL BE MADE NET THIRTY (30)
                                       ATTENTION ACCOUNTS PAYABLE, UNLESS A   DAYS AFTER RECEIPT OF A CORRECT INVOICE,
                                       PURCHASE CONTRACT NOTE REQUESTS INVOICE   UNLESS OTHERWISE AGREED TO IN THE TERMS AND
                                       SUBMITTAL TO ANOTHER ADDRESS.   CONDITIONS OF THIS PURCHASE CONTRACT AND/OR
                                                                                     PURCHASE CONTRACT CHANGE.

•••

PAGE: 1 OF 5   PURCHASE CONTRACT NO: B 233234

Fig. 4-6. Company purchase contract.

PURCHASE CONTRACT NO: B 233234   CHANGE NO: 1   PURCHASE CONTRACT   PAGE: 2 OF 5

ITEM NO	SUB ITEM NO	ORDER QUANTITY	UM	DESCRIPTION	SCHEDULES SHIP	SCHEDULES QUANTITY	BILLING QUANTITY	UM	UNIT PRICE	EXTENDED PRICE
1		1	EA	CONTROL ROOM, INCLUDING INSTALLATION AND DRAWINGS, W/C AIR CONDITIONING, AS PER SPECIFICATION E 1207 DATED 06/12/87			1	EA	$56,936.00	$56,936.00
					08/03/87	1				
2		1	LT	VINYL TILE FLOOR INSTALLED ON FIRST FLOOR OF ITEM 1			1	LT	$1,457.00	$1,457.00
					10/05/87	1				

NET PURCHASE CONTRACT VALUE $58,393.00

NOTES

1   CONFIRMING ORDER ENTRY 06/30/87 BY TELECON, BETWEEN

DO NOT DUPLICATE.

2   INVOICES FOR SERVICES OR PARTS DELIVERED AGAINST THIS PURCHASE ORDER SHALL BE MAILED DIRECTLY TO:

ATTENTION:

3   D & A ENGINEERING QUOTATION DATED 06/01/87, IN RESPONSE TO   REQUEST FOR QUOTE NO. 6360   , IS INCOR- PORATED IN THIS ORDER. SHOULD CONFLICT ARISE, THE TERMS AND CONDITIONS AND ACCOMPANYING DOCUMENTS OF THIS ORDER SHALL PREVAIL.

4   WARRANTY SHALL BE FOR A MINIMUM PERIOD OF ONE YEAR FROM DATE OF FINAL ACCEPTANCE, DURING WHICH PERIOD THE MANU- FACTURER SHALL BE RESPONSIBLE FOR REPLACEMENT OR REPAIR OF ANY DEFECTIVE PARTS, COMPONENTS, OR SYSTEMS, THE FAILURE OF WHICH IS DUE TO FAULTY MATERIAL, WORKMANSHIP OR ENGINEERING.

5   ALL CHANGES TO THIS ORDER AND ACCOMPANYING DOCUMENTS MUST HAVE THE APPROVAL OF

PURCHASE CONTRACT NO: B 233234   CHANGE NO: 1   PURCHASE CONTRACT   PAGE: 3 OF 5
NOTES

6   SHIPMENTS TO   DIVISION MUST REFERENCE THE PURCHASE ORDER ON ALL APPLICABLE FREIGHT BILLS OF LADING, WHETHER PREPAID OR COLLECT.

7   A WRITTEN ACKNOWLEDGEMENT OF THIS PURCHASE ORDER IS REQUIRED TO BE RETURNED TO THE BUYER NO LATER THAN FIFTEEN (15) DAYS AFTER RECEIPT OF ORDER.

8   SUPPLIERS PERFORMING CONTRACTED ACTIVITIES ON OWNED OR LEASED PREMISES AND/OR GOVERNMENT PROPERTY IN SUPPORT OF THIS ORDER, I.E. INSTALLATIONS, OPERATIONS, DEMONSTRATIONS, MAINTENANCE, OVERHAUL, INSPECTION, TRAINING, RELOCATION AND/OR REMOVAL OF RENTED/LEASED EQUIPMENT, ARE REQUIRED TO PROVIDE THE BUYER WITH CERTI- FICATION OF THE FOLLOWING INSURANCE REQUIREMENTS.

SELLER'S RELATIONSHIP TO BUYER IN THE PERFORMANCE OF THIS ORDER IS THAT OF AN INDEPENDENT CONTRACTOR. THE PERSONNEL PERFORMING SERVICES UNDER THIS ORDER SHALL AT ALL TIMES BE UNDER SELLER'S EXCLU- SIVE DIRECTION AND CONTROL AND SHALL BE EMPLOYEES OF SELLER AND NOT EMPLOYEES OF BUYER. SELLER SHALL PAY ALL WAGES, SALARIES AND OTHER AMOUNTS DUE ITS EM- PLOYEES IN CONNECTION WITH THIS ORDER AND SHALL BE RESPONSIBLE FOR ALL REPORTS AND OBLIGATIONS RE- SPECTING THEM RELATING TO SOCIAL SECURITY, INCOME TAX WITHHOLDING, UNEMPLOYMENT COMPENSATION, WORK- MEN'S COMPENSATION AND SIMILAR MATTERS.

SELLER SHALL COVER OR INSURE ALL OF ITS EMPLOYEES PERFORMING SERVICES UNDER THIS ORDER IN COMPLI- ANCE WITH THE APPLICABLE LAWS RELATING TO WORK- MEN'S COMPENSATION OR EMPLOYER'S LIABILITY INSUR- ANCE.

SELLER SHALL INDEMNIFY AND SAVE HARMLESS BUYER AND THE UNITED STATES OF AMERICA FROM AND AGAINST ALL CLAIMS FOR (1) BODILY INJURIES, INCLUDING DEATH, OR DAMAGE TO PROPERTY CAUSED BY ANY NEGLIGENT ACT OR OMMISSION OF SELLER OR ITS EMPLOYEES IN CON- NECTION WITH THE PERFORMANCE OF THIS ORDER, OR (2) BODILY INJURIES, INCLUDING DEATH, TO SELLER'S EMPLOYEES CAUSED BY THE CONDITION OF ANY PREMISES, EQUIPMENT OR OTHER PROPERTY BEING USED OR OPERATED BY ANY OF SELLER'S EMPLOYEES IN

NOTES

CONNECTION WITH PERFORMANCE OF THIS ORDER.

IF SERVICES REQUIRE PERFORMANCE ON THE
COMPANY PROPERTY, OR IF DELIVERY OF MATERIALS BY
SELLER'S PERSONNEL IS TO AREAS OTHER THAN THE
ASSIGNED RECEIVING AREAS, SELLER WILL SECURE AND
MAINTAIN SUCH COMPLETE AUTOMOBILE AND GENERAL
LIABILITY INSURANCE AS WILL PROTECT SELLER AGAINST
ALL CLAIMS FOR BODILY INJURY, INCLUDING DEATH, AS
WELL AS ALL CLAIMS FOR PROPERTY DAMAGE WHICH MAY
ARISE FROM OPERATIONS UNDER THIS ORDER, WHETHER
SUCH OPERATIONS ARE BY SELLER OR ITS SUBCONTRACTOR,
OR THEIR RESPECTIVE AGENTS OR EMPLOYEES. SUCH
INSURANCE SHALL BE OBTAINED IN NOT LESS THAN THE
FOLLOWING MINIMUM AMOUNTS:

TYPE	AMOUNT
A. WORKER'S COMPENSATION	STATUTORY LIMITS
EMPLOYER'S LIABILITY	$500,000
COMPREHENSIVE GENERAL LIA-BILITY INCLUDING CONTRACTUAL.	
BODILY INJURY	$1,000,000 EACH OCCURRENCE
PROPERTY DAMAGE OR	$1,000,000 EACH OCCURRENCE
COMBINED SINGLE LIMIT	$1,000,000 BODILY INJURY AND PRO-PERTY DAMAGE
COMPREHENSIVE AUTOMOBILE LIABILITY.	
BODILY INJURY	$1,000,000 PER PERSON OR PER ACCIDENT
PROPERTY DAMAGE OR $1000,000 COMBINED SINGLE LIMIT BODILY INJURY AND PROPERTY DAMAGE.	$1,000,000 PER ACCIDENT

B. PRIOR TO THE COMMENCEMENT OF WORK HEREUNDER,
SELLER SHALL FURNISH TO BUYER A CERTIFICATE OF
THE ABOVE REQUIRED INSURANCE. THE POLICIES
EVIDENCING REQUIRED INSURANCE SHALL CONTAIN AN

NOTES

ENDORSEMENT TO THE EFFECT THAT CANCELLATONS OR
OR ANY MATERIAL CHANGE IN THE POLICIES ADVERSELY
AFFECTING THE INTERESTS OF BUYER OR THE GOVERN-
MENT IN SUCH INSURANCE SHALL NOT BE EFFECTIVE
UNTIL 30 DAYS AFTER WRITTEN NOTICE THEREOF TO
BUYER.

C. SELLER AGREES TO INSERT THE SUBSTANCE OF THIS
CLAUSE, INCLUDING THIS PARAGRAPH C., IN ALL SUB-
CONTRACTS HEREUNDER.

D. THE INDICATION OF MINIMUM INSURANCE COVERAGE
LIMITS DOES NOT ACT IN ANY WAY TO LIMIT THE LIA-
BILITY OF THE SUPPLIER.

9   POC NO. 1 ADDS ITEM 2 PER QUOTATION DATED 9/24/87.

Fig. 4-6. Continued.

The NOTES section of the purchase contract is important because it contains everything else of importance regarding the performance and responsibilities, etc., of the supplier and the purchaser.

Purchase contract notes

- A confirming order entry date, by method (telephone, telegram, letter).
- Company (XYZ Company), buyer's name, and supplier's contact (name).
- XYZ Company's Accounts Payable address.

- Incorporation of supplier's quotation (with date) as a part of the purchase contract.
- Caveat that, should a conflict arise between the quotation and the purchase contract, the terms and conditions of the purchase contract shall prevail.
- Warranty period spelled out as being one year from the date of final acceptance.
- The supplier is held responsible for the repair or replacement of any defective parts, components, or systems whose failure is due to faulty material, workmanship, or engineering.
- All changes to the purchase order (purchase contract) and accompanying documents must have the approval of the XYZ Company's buyer, named in the purchase contract.
- Shipments to the XYZ Company must reference the purchase order on all applicable freight bills of lading, whether prepaid or not.
- A written acknowledgment of the purchase order is to be returned to the buyer of the XYZ Company within fifteen days of receipt of the order.
- The supplier must provide the buyer of the XYZ Company certification of insurance requirements when performing contracted activities on the XYZ Company's premises (whether or not the premises are owned or leased by the XYZ Company makes no difference), in complying with the purchase order in the following areas:
  - Installation
  - Operations
  - Demonstrations
  - Maintenance
  - Overhaul
  - Inspection
  - Training
  - Relocation and/or removal of rented or leased equipment

- Certification of insurance requirements:
  - The seller's relationship to the buyer in the performance of the purchase contract is that of an independent contractor.
  - The personnel performing services under this purchase contract shall, at all times, be under the seller's exclusive direction and control, and shall be employees of the seller and not employees of the buyer.
  - The seller shall pay all wages, salaries, and other amounts due its employees in connection with this order, and shall be responsible for all reports and obligations respecting them relating to social security, income tax withholding, unemployment compensation, workmen's compensation, and similar matters.
  - The seller shall cover or insure all of its employees performing services under this order in compliance with the applicable laws relating to workmen's compensation or employer's liability insurance.
  - The seller shall indemnify and save harmless buyer and the United States of America from and against all claims for:
    - ~Bodily injuries, including death, or damage to property caused by any neg-

ligent act or omission of seller or its employees in connection with the performance of this order; or

~ Bodily injuries, including death, to seller's employees caused by the condition of any premises, equipment, or other property being used or operated by any of seller's employees in connection with the performance of this order.

○ If services require performance on the XYZ Company property, or if delivery of materials by seller's personnel is to areas other than the assigned receiving areas, seller will secure and maintain such complete automobile and general liability insurance as will protect seller against all claims for bodily injury, including death, as well as any claims for property damage that may arise from operations under this order whether such operations are by the seller or its subcontractor, or their respective agents or employees. Such insurance shall be obtained in not less than the following minimum amounts:

Type	Amount
A. Worker's compensation Employer's liability 1. Comprehensive general liability, including contractual.	Statutory limits $500,000
Bodily injury Property damage, or	$1,000,000 each occurrence $1,000,000 each occurrence
Combined single limit	$1,000,000 bodily injury and property damage
2. Comprehensive automobile liability	
Bodily injury	$1,000,000 per person, or per accident
Property damage or $1,000,000 combined single-limit bodily injury and property damage	$1,000,000 per accident

B. Prior to the commencement of work hereunder, the seller shall furnish to buyer a certificate of the required insurance (in other words, the seller must prove to the buyer's satisfaction that the insurance is in force). The policies evidencing required insurance shall contain an endorsement to the effect that cancellations or any material change in the policies adversely affecting the interests of the buyer or the government in such insurance shall not be effective until 30 days after written notice thereof to the buyer. (In effect, this gives the buyer 30 days in which to make other arrangements, or to take action to mitigate any adverse effects that might arise from cancellation of the seller's insurance policies.)

C. The seller agrees to insert the substance of this clause, including this paragraph C, in all subcontracts hereunder.

D. The indication of minimum insurance coverage limits does not act in any way to limit the liability of the supplier.

Note that some of the smaller suppliers might be inhibited from bidding because of the difficulty in obtaining various forms of insurance, especially insurance concerning liability, since this type of insurance has become very expensive due to the large awards made as the result of the lawsuits. In my experience, however, it is possible to obtain liability limits of $1 million or more for only the period of time necessary to cover the work on the buyer's premises. For example, if it is estimated that the work will take three months, then an insurance underwriter is capable of prorating the cost of such policies for 3/12 year, thus cutting the annual premium to one-fourth the normal amount.

Although the XYZ Company is the buying company of our example, the Endure-A-Lifetime Products Company, Inc. is the selling company. The last page of the proposal has the following paragraphs:

NOTE: Pre-paid & add freight will have a 10% of freight handling charge added.

We refer you to the reverse side of this quotation, specifically 11, 16, and 17.

Factory shipment (days) after receipt of order and necessary approvals.
TERMS: 1% 10 days, net 30 after credit approval.

ENDURE-A-LIFETIME PRODUCTS, INC.

/S/ *John Doe*
_____
JOHN DOE (Sales Manager)

This quotation, attached provisions, and Form SL-5113 to become a valid purchase order to ENDURE-A-LIFETIME PRODUCTS, INC. when accepted and signed by the purchaser and approved at Miami, Florida, by an officer of ENDURE-A-LIFETIME PRODUCTS, INC.

Accepted by: Approved by:

_____   _____

On the reverse side of each page of the quotation, are found the following provisions:)

Provisions

1. Endure-A-Lifetime Products, Inc. or its subsidiaries, hereinafter called the Company, hereby agrees to deliver the items described (contract merchandise) in accordance with terms and conditions herein.

2. This proposal becomes the contract when accepted by the purchaser and approved in writing by an authorized officer of the Company at Miami, Florida. Regardless of where executed geographically, the parties agree that this contract has been entered into in Dade County, Florida, and shall be governed by the laws of the State of Florida, both as to interpretation and performance. Any action at law, suit in equity, or other judicial proceeding for the interpretation or enforcement of this contract or any provision thereof shall be instituted only in courts of competent jurisdiction in Dade County, Florida.

3. All payments provided by this contract shall be payable at the Company's office in Dade County, Florida. If payment as provided in this Contract is not made in full and is placed with an attorney for collection, the purchaser agrees to pay all costs of collection, including court costs and reasonable attorney's fees. Deferred payments and/or past-due amounts to bear interest at the maximum allowable rate.

4. The purchaser acknowledges that the merchandise is being manufactured to its order in accordance with its special request—i.e., specifications and/or drawings that the purchaser has sent to Endure-A-Lifetime Products, Inc. of the required Contract Merchandise. Shop drawings requested by purchaser must be approved in writing by the purchaser.

   In consideration of ten dollars and other valuable consideration, the receipt of which is hereby acknowledged, the Purchaser agrees to indemnify the Company and save it harmless against any and all claims, judgments, decrees, costs, expenses (including reasonable attorney's fees) or any other loss which the Company might sustain by reason of the manufacture of the merchandise called for by the Contract according to the plans, drawings, and specifications furnished to the Company by the Purchaser, and/or according to the shop drawings approved by the Purchaser. The Purchaser convenants that it will upon request of the Company and at Purchaser's expense defend or assist in the defense of any suit or action which may be brought against the Company by reason of the manufacture of the merchandise.

5. In the event of any liability on the part of the Company arising out of the design, manufacture, sale, or use of the merchandise sold hereunder, not subject to Purchaser's indemnification contained herein, the Company's liability shall be coextensive with and shall not exceed its liability insurance coverage.

6. The Company is not responsible for its inability to fulfill or complete the Contract because of governmental laws or regulations, strikes, transportation delays, shortage of material, government allocations or restriction, or any other cause or catastrophe over which it has no control.

7. It is agreed that there are no understandings, agreements, representations, or warranties, oral or written, express or implied, except those contained herein, and this instrument contains the entire agreement between the parties. No subsequence amendment or modification of this Contract shall be effective unless in writing and signed by both parties.

8. The procurement of necessary building permits shall be the responsibility of the Purchaser. Failure of the Purchaser to secure such permits shall not relieve Purchaser of any of its obligations created by any of the provisions of this Contract.

9. If the Contract calls for installation of the material, the Purchaser shall pay to the seller upon delivery and prior to installation 90% of the purchase price with balance due upon completion of installation. This shall apply unless otherwise noted in Contract. If installation is outside Dade County, Florida, boundary lines, this intermediate payment shall be made prior to contract merchandise leaving Dade County, Florida.

10. The Purchaser does understand that installation and/or delivery dates, if given, must be approximate because of the nature of our business. The Purchaser cannot cancel this Contract for failure to deliver or install on the approximate delivery date. The Company shall have the right and the Purchaser must accept delivery and/or installation as soon after such date as the seller can, with reasonable diligence, deliver and/or install the Contract merchandise.

11. Delivery subject to credit approval notwithstanding any proposed shipping dates.

12. The Purchaser acknowledges that the merchandise is being made up in accordance with his special request and to his order, and as a result this order is not subject to cancellation or countermand. In the event the Purchaser attempts to countermand or cancel the same or refuses to accept said merchandise when completed, then the total amount of the Contract shall immediately become due and payable; provided, however, that if merchandise is not actually in possession of Purchaser, then and in that event a credit shall be given to the Purchaser for the salvage value of such merchandise (as determined by Company practice), which salvage value shall be deducted from Contract price and balance of Contract price shall immediately become due and payable.

13. The Company will not be responsible for any contingent liabilities arising out of third-party contracts.

14. FOB Factory. All prices are FOB Factory unless specially stated otherwise herein. The customer has fourteen (14) days after receipt of any of the Contract merchandise to report any items damaged, missing, or lost in transit. After a receipt of a copy of the freight claim, all items will be replaced, all charges for Purchaser's account. All freight claims must be made by Purchaser. Delivery of all materials shall be considered complete when the Company delivers to any carrier for transshipment to delivery place.

15. All prices quoted are U.S. dollars. Any checks tendered for payment will be accepted at par only. All charges for purchaser's account, import or export charges, port fees, taxes, or any other charges not specifically included in quotation will be for Purchaser's account.

16. Terms quoted are tentative only. Final terms will be determined at order acceptance by Company.
17. Taxes. Taxes now or hereafter levied by Federal, State, or Local authorities upon sale of this material are not included in this contract unless specifically mentioned.
18. Back Charges. No back charges will be allowed unless approved beforehand by the Company in writing. Back charges not made within 30 days of delivery will not be considered.
19. Changes. No commitments to alter or change the foregoing provisions will be binding on the Company unless such changes are requested in writing by authorized persons of the Purchaser, and confirmed in writing by an Officer of the Company.
20. If and to the extent any provisions of the Contract shall be held by any law or court of competent jurisdiction to be in whole or in part invalid or unenforceable, such provision or part thereof shall be deemed to be surplusage and, to the extent not so determined to be invalid or unenforceable, each provision shall remain in full force and effect.

The form SL-5113, referenced in these paragraphs, indicates the product or "building finish" tolerances, as follows:

Building finish tolerances

Tolerances: Any surface irregularity shall be considered as suitable for use if such irregularity is not noted from a distance of 15 feet. If during the fabrication process traffic marks or process marks are noticed, they may be repainted to meet this 15-foot visual criteria.

All mill finish aluminum will be received from the mill. Surfaces could be varying shades, but no waterstains or crazing on exposed surfaces will be shipped.

Prepainted finish on aluminum or steel shall be material received by Endure-A-Lifetime Products, Inc. from our suppliers. It shall have a uniform surface color. Smudges, drips, alligatoring, etc., shall not be acceptable. There may be high spots on the sheet that have been rubbed and lost color.

Shop-painted finish shall be material fabricated by Endure-A-Lifetime Products, Inc., and painted in our paint department. Mill finish material will be cleaned and etched in proprietary chemical solutions, and then prepared to receive paint. Repainted finish shall be material previously painted by our supplier. It will be cleaned and prepared to receive paint.

Paint shall be commercial quality as furnished to Endure-A-Lifetime Products, Inc. by our supplier. It shall have properties as needed for adhesion, gloss, and color retention.

Due to the nature of finishing materials, there shall be no warranty of color or gloss. Weathering will only be warranted as to that which the paint manufacturer shall make.

The provisions which are singled out and emphasized on the last page of the quotation are paragraphs 11, 16, and 17.

Para. 11 Paragraph 11 states that delivery is subject to credit approval notwithstanding any proposed shipping dates.

Para. 16 Paragraph 16 indicates that the terms quoted are tentative only, and that final terms will be determined at order acceptance by the Endure-A-Lifetime Products Company.

Para. 17 Paragraph 17 concerns the payment of taxes, and that any taxes levied by the Federal, State, or Local Authorities upon sale of the material of the Endure-A-Lifetime Products Company are not included in the contract unless specifically mentioned.

As you will notice, there is a good deal of boiler-plate, or standard, clauses in both the purchase contract issued by the XYZ Company and in the "provisions" section of the Endure-A-Lifetime Products Co. quotation. There is a tendency, therefore, that apparent redundancy will occur in other purchase contracts and quotations, the former by the buyer and the latter by the supplier. It really makes no difference in the final outcome, inasmuch as both companies are signatory to either of the documents.

What constitutes an order and acceptance It is commonly stated that a contract is only as good as the parties signing the agreement. Verbal agreements are no exception to this adage, and under common law, all verbal agreements between authorized parties are perfectly binding. However, the Uniform Commercial Code indicates that a contract of $500 or over must be in writing to be enforceable. Generally speaking, put everything in writing, including any amendments and revisions to each contract and ask for acknowledgments.

In most instances, a contract does not exist until the seller makes an offer and the buyer accepts the offer. Under common law, the officer and acceptance must agree in every detail. If a buyer issues a purchase order that is different from the terms of the offer, under common law it is problematical that a contract existed; however, under the Uniform Commercial Code, a contract would exist as long as both the buyer and the seller acted as though there were a contract.

In some circumstances, the price that the seller is willing to sell his merchandise for is limited by the length of time it takes the buyer to react. For example, a quotation should be accompanied by a statement that the prices quoted are good for only so many days (or months) after the date of the quotation. Difficulties might arise between the buyer and the seller if an unreasonable length of time has elapsed between the date of the quotation and the buyer's acceptance. It appears unreasonable that a year after the quotation date, the buyer can crank out a purchase order based on the quoted price at that time. A reasonable amount of time might be assumed to be approximately 90 days. Lawsuits, however, have arisen out of even less substantive matters than what constitutes a reasonable period of time.

When the buyer and the seller both agree to the terms and conditions of the contract without exception, there does not appear to be any cause for legal difficulty. The buyer, in every instance, should review the seller's quotation in detail. If there are any terms and conditions that appear unsatisfactory, the buyer should make them known to the seller in writing or the buyer may include a stipulation that the acceptance of the seller's offer is conditional and limited by the terms and provisions of the buyer's purchase order.

The materials management concept, encompassing as it does the entire process of the movement and flow of materials from the supplier to and through manufacturing, storage, and distribution operations, depends heavily on the purchasing department to ensure that materials are received according to schedule, in the proper quantities, in acceptable condition (a quality problem), and packaged (or not packaged, depending on the end use of the product) in such a manner that there is very little wasted effort or packaging debris to be removed. Thus it is that the purchase order specifies, in detail, how the material is to be shipped to the company so that it arrives at the plant in the best possible shape to be used—at the earliest possible time with the minimum of storage or handling. The purchasing department has an enormous responsibility in the overall materials management concept, and it is the linchpin that cements the process of moving materials in the entire manufacturing or distribution chain.

Canceling a purchase order   Since the delivery of materials to the company is a crucial factor in the materials management program, a seller's proposal that includes a clause permitting the "cancellation of a purchase order due to circumstances beyond our control," such as a disastrous fire, labor strike, or the like, is quite common. It is useless to depend on a general clause, usually printed on the reverse side of the purchase order, that directs the supplier to deliver the materials within a specified time frame and makes the delivery an essential part of the purchase contract; however, it is necessary to have this clause as a matter of course. It stands to reason that this will shift the burden of proof to the seller to show why the seller should not be liable for damages caused to the buyer because of the seller's inability to deliver the materials as ordered.

One company, at which I was employed, was a very heavy purchaser of steel. As you know, the steel industry, like most other capital-intensive industries, such as the automobile industry, have three-year labor contracts with their respective unions. Thus, at three-year intervals, there was heavy purchasing volume of steel products prior to the expiration of the steel companies' labor contracts. This was a strike-hedge, which had considerable impact on prices and space, because our company was not alone in their anticipation of the possible discontinuation of steel production. Another influence, or impact, this philosophy engendered was the shift to off-shore purchasing of steel from Europe and the Orient.

The effect of pricing   The price of materials has a definite effect on the flow of materials at certain price levels. For example, some purchasing departments buy materials and products on verbal orders, usually following up a call to a supplier with a written purchase order. In other instances, however, the buyer will simply follow up the verbal order with a handwritten purchase memo.

In some small companies, the arrangements are sometimes so loose that no paperwork is produced by the buyer's company until the vendor presents an invoice for payment. The buyer usually receives a copy of the invoice and it is his or her job to verify that the merchandise has actually been received and is of acceptable quality, and that the quantity is correct. The buyer, generally, has a fairly good idea of the value of the product received, especially if the company has a record of prior purchases.

Nevertheless, this practice, although usually restricted to small dollar-value purchases, is not a recommended method. One of the reasons is that if the price appears to be out of line, the buyer is forced either to negotiate the difference with the vendor or to return the merchandise for credit and then risk being charged with a restocking cost. The additional paperwork and trouble that these rare instances generate makes the regimen of issuing a purchase order for each purchase, wherein the terms and conditions of the order are clearly and unequivocally stated, very much an accepted practice especially in all larger, or very well run, companies.

Problems in pricing sometimes occur when the delivery of the material takes place at an extended period of time after the order has been placed. Inflationary impacts of labor, materials, and overhead costs, in general, might change during this interval so that the supplier who is doing the fabricating, or the producer of the raw material, might find that his or her costs are burgeoning, thereby shrinking his or her profit margin. The reverse is also true in a time of declining prices, which is rare in this decade, except for certain commodities such as metals, which fluctuate at a very rapid rate on occasion.

In instances where there are relatively sharp increases or decreases in the cost of production, either the buyer or the seller might desire to renegotiate the prices that have been agreed upon in the purchase contract. Unless provisions in the purchase contract permit change, both the seller and buyer are legally bound to agree on the prices and quantities contained in this document.

Some companies use or accept purchase contracts for the delivery of materials or products over an extended period of time without any provisions regarding changes in quantities or prices, simply because they feel that these things can be negotiated, if the need arises. Where companies have been doing business with each other for a long period of time, such changes usually can be made amicably. Or, if the companies involved are well established, with good reputations, they might feel that their public image is more important than a few pennies in price. You are fortunate, indeed, when dealing with this type of company. As the saying goes, *caveat emptor*.

Although there are some companies that have very carefully phrased purchase orders that include provisions for price changes, almost all of the federal, state, and local government purchase contracts contain clauses referring to price changes, which are dictated by their purchasing regulations. The same caution applies to most governmental suppliers who have prime contracts and have to do business with subcontractors and other suppliers.

While most governmental and prime subcontractors include standard provisions that are circumscribed by regulation, other companies and purchasing authorities have greater leeway; thus, the terms and conditions they might impose vary considerably, depending on the industry segment they fall in and the general receptivity of their suppliers. Several classifications may determine how price changes can be derived:

- Price redetermination
- Cost plus
- Time and material
- Price escalation
- Open-end pricing

Price redetermination   This type of contractural clause permits periodic renegotiation of the contract price. It is a much more lenient view of pricing than anticipated by the more restrictive "escalation" clause since the price does not have to be related directly to any changes in wage rates or raw material, or increases in product cost.

The price redetermination clause generally permits the price of an item to be changed with some sort of periodic frequency, semi-annually, or annually in order to take into consideration such things as changes in basic costs or the supplier's experience in producing the product. Price redetermination clauses have their most effective application in long-term contracts that extend over a relatively long period of time, or in cases where neither the buyer nor the seller has had very much experience with the product, either in buying it or in its production.

Price redetermination has its largest application in purchasing by the military establishment; however, there are also useful civilian applications of this methodology. For example, large mail-order houses might decide to buy a certain item from a supplier using a long-term contract in order to ensure a definite supply of the item and excellent—i.e., competitive—pricing. It knows that the supplier also must make a profit to stay in business. In this instance, the profitability figure for the supplier might be 10 percent of the gross selling price. Any cost savings due to higher quantity, lower costs, cheaper materials, etc., will be passed on to the buying company in terms of a periodic price review. Additional legal phraseology, in this instance, would have to be added to the purchase contract, which would permit the buyer to audit costs, etc.

In the final analysis, every contract is subject to renegotiation, and primarily where pricing is concerned, simply because the terms of a contract must be at least more favorable than the monetary penalties for nonperformance. In my experience, a large midwestern company manufacturing agricultural implements had entered into a contractual arrangement with a small box-making concern in order to ensure a supply of shipping containers. A fixed price per container was agreed upon and a purchase contract was issued based on a certain quantity of containers to be shipped each month. The small box manufacturer's pricing was based on the costs to be generated by this expected volume. Unfortunately, the quantities that were to be shipped each month depended on the larger company's requirements, which occasionally fell much below the anticipated volume. Since the box maker's costs depended on a break-even point that was only slightly below the anticipated volume, the unit cost per container had to be renegotiated.

Cost plus Sometimes neither the buyer nor the seller has a very precise knowledge of the cost of a product that must be fabricated, possibly for the first time. For this reason, in all fairness to the seller (fabricator), the buyer will venture into a cost-plus contract with the manufacturer, who has some expertise in the area, but who has not, until this time, produced a product of similar characteristics.

The seller—in this instance, the manufacturer—agrees to manufacture the item(s) at his cost, which is a price per hour that includes the manufacturer's wage rates plus a small profit margin on the cost of materials and wages.

Much of the military establishment's purchasing in the research and development area takes this form because of the relatively experimental nature of the work involved.

When contracts such as the cost-plus variety are entered into, there are at least two elements that must be introduced by the buyer:

- Auditing of the manufacturer's books for the project should be allowed
- The buyer's resident engineer or resident inspector should be permitted on the manufacturer's premises with access to every component of the project from hardware to accounting.

The main reason for these precautions is that they will eliminate, or minimize, wasteful practices or padding of the company's books to increase the margin of profit. Many cost-plus contracts have a fixed-fee appendage that permits the supplier to recover his costs and a fixed amount over and above his costs. The fixed fee might be as much as 10 percent of aggregate cost but, of course, it would vary with the scope of the project.

Time and Materials In order to control the total cost of a purchased item(s), a buyer will enter into a time and materials contract with a supplier. The buyer might also opt for this kind of contract when the scope of the project is of relatively short duration. The buyer and seller agree on a reasonable labor cost, and then the costs of material used to produce the product are added.

If the supplier is especially astute, he or she will attempt to add a handling markup on the material costs in order to recover the overheads on the purchasing and record-keeping functions that will be necessary in order to administer the contract. If extensive engineering time is included in the project, the seller probably will charge the buyer at least 2 to $2^1/_2$ times the average hourly rate that the engineering staff earns in order to ensure a profit and to cover all of the fringe benefits of the sellers' employees. Shop time also might be included in the contract at several times the prevailing wage for the same reason.

Although cost-plus and time and materials contracts are relatively difficult to administer and are fraught with the possibilities of a certain amount of padding in order to increase the seller's profits, some tasks are so nebulous that sometimes that is the only possible way to obtain a contractor. Where necessity demands this type of purchase contract and where a fixed-price contract is not possible because of all the hidden problems that might be unearthed as a supplier progresses into the development of the item, the scope of the project will determine how much surveillance is required to assure the buyer that he or she is getting what is being paid for. A fixed-price contract would, under these circumstances, be so heavily loaded with contingencies that the buyer would have to use the time and materials approach, however apprehensive he or she might be of the outcome.

Escalation Escalation clauses in contracts, especially long-term contracts of a year or more in duration, permit the realignment of cost and profit so that the arithmetic ratio between the two is compatible with the original agreement, as nearly as possible. The buyer and seller both have a stake in any escalation adjustment because the buyer wants to obtain the best possible price for the item, and the seller wants to maintain his or her profit margin despite higher costs. If there is unlimited escalation, the buyer will lose all the advantage of entering into a long-term contract.

Most of the time-escalation clauses govern the changes that might take place in either labor or material costs, and often in both of these ingredients. Sometimes, however—and this is becoming more common—the escalation is linked to some public index, such as the Consumers' Price Index, the Eleventh District of the Federal Reserve Board's Cost of Money Index, or some other locally available information, such as the Bureau of Labor Statistics' (BLS) wage guidepost.

An important factor in price-escalation clauses is that there should be an absolutely clear definition of what type or method is to be used in determining the escalation factor. It is a mistake on the part of the buyer to base escalation on the seller's internal costs because this could become a source for dispute. To prevent misunderstanding, the BLS wage index for the seller's industry segment could be used, whereby every five-cent increase in hourly wage could cause a ten-cent increase in the contract price after every quarterly review of the contract. The escalation clause should provide an upper limit to protect the buyer, such that at no time will the escalation exceed a certain amount, say 10 percent of the original contract price.

Open-end price The volatility of market prices for certain commodities often makes it impractical for both buyers and sellers to commit themselves to a fixed price for any great length of time. Therefore, an open-end contract will ensure a supply of the commodity at prevailing prices—at the time of purchase. When this type of contract is entered upon, the prevailing price might be quoted as a reference source, but the supplier might charge the buyer the price in effect at the time of delivery.

Most open-end price contracts are employed for the basic raw materials of industry, whose prices are fixed in the trading pits of the world's great commodity markets. Producers of such raw materials as lumber, aluminum, and steel, who do not rely on the commodity markets for their pricing, do not sell their materials any other way, but use the price in effect at the time of delivery to exact their pound of flesh from the buyer. Knowing this fact, and trying to obtain the best prices possible, buyers will shop around to obtain competitive prices. Buyers tend to shy away from open-end pricing because it can become too costly.

Other purchasing factors affecting materials management

In the complex arena of purchasing, there are many terms and phrases of significance, for example, *F.O.B., 1% net 30, warranty*, and *specifications*. So that the materials manager can use these terms to his or her advantage, a certain familiarity with the more mundane aspects of purchasing is in order.

F.O.B. This is the abbreviation of *free on board*, sometimes vulgarized as "freight on board," but it means the same thing. For example, if the buyer indicates that the material being purchased is *F.O.B. our plant*, then the buyer would take title to the merchandise when it is placed on the delivery dock at the buying company's plant.

The useful thing about F.O.B. our plant is that the supplier is responsible for the transportation cost, and if there is any damage to the material, it becomes a problem for only the seller and the transporter. The seller must file a damage claim against the carrier in that event, relieving the buyer of any responsibility in that area.

The F.O.B. designation not only specifies who pays the freight charges, but also indicates when title for the material changes hands. For example, when the *F.O.B. shipping point* is the seller's plant, then the buyer takes title when the seller places the material into a common carrier. The buyer then pays the carrier directly, usually within fifteen days of being presented with the charges by the carrier. If there are any damages to the freight, the buyer must present and negotiate the damage claims directly with the carrier. If the buying company has a sufficiently large volume of freight both inbound and outbound, it usually will have a traffic department that will prefer to make the F.O.B. point the seller's plant because in most instances it will be able to negotiate better freight rates if it has complete control of the merchandise at the seller's plant.

Freight equalization   Freight equalization occurs in an *oligopolistic* pricing market; that is, where there are few sellers, such as the steel and aluminum industries, and where the seller is anxious to compete in an area in which a competitor's plant may have a freight advantage. As an example, if the freight rate from the competitor's plant to the customer is $1.00 per hundred weight (cwt.) and $2.00/cwt. from the seller's plant, then the seller might give the customer a freight equalization of $1.00/cwt. to compensate for the difference; in effect this amounts to a freight rebate.

The principal reason for making the change in the freight charge, rather than reducing the price of the commodity, is that the seller does not want to reduce prices in view of the local market that it supplies. Thus, the buyer would pay the full freight charge, but would deduct the $1.00/cwt. from its invoice. The buyer's purchase order would indicate that the price was F.O.B. shipping point and that freight equalization adjustment is to be allowed.

Export shipments   In this era of increased global commerce, there are two frequently used terms: *free alongside ship* (f.a.s.) and *cost-insurance-freight* (c.i.f.). Free alongside ship is used to denote the port of export, for example, "f.a.s. Hong Kong." The buyer must indicate the port, berth—i.e., dockspace—and vessel. It is the supplier's responsibility, in this instance, to see that the merchandise makes the sailing and must then provide the buyer with a receipt indicating that the merchandise was delivered to the ship's loading crew undamaged; this is the point at which the buyer takes title to the merchandise.

Cost-insurance-freight is closely akin to the F.O.B. our plant because the buyer designates the port to which the material is to be sent. As an example, if a U.S. buyer purchases material in Taiwan, it can require that the goods be sent "c.i.f. Seattle," and the supplier assumes responsibility for the material up to the point that the vessel (or airplane) arrives in Seattle. At the point of docking, it is the buyer's turn to assume responsibility and take title to the material; thus he or she must oversee that the material is unloaded and arranges for delivery to the final leg of its destination in the United States.

The abbreviated term *c. and f.* stands for cost and freight, without the insurance. Under this arrangement, the vendor does not assume any liability if the vessel carrying the material founders en route to the United States.

The negative side of purchasing that affects materials management

In a less than perfect world, purchased materials sometimes do not meet the buyer's specification or for various other legitimate reasons must be returned to the supplier either for credit or for replacement. Sometimes, for other equally valid reasons, purchasing contracts must be canceled. In many states regulations governing these actions are fairly explicit, and merchandise may be returned to the vendor, provided he or she is notified expeditiously after the buyer's receipt of the material.

Warranties In every instance, the buyer may perform reasonable tests on the materials that have been purchased. One of the tenets of materials management is that "the right quantity of materials, of the right quality, be supplied at the right place, at the right time." Therefore, the purchasing department must depend heavily on the Quality Control (Q.C.) organization of the company to ascertain that a quality product— i.e., one conforming to the company's specifications—has been received. If the material (component part, etc.) is not of sufficient quality, the Q.C. department places a "hold" on the shipment and notifies the buyer immediately. The buyer promptly notifies the vendor of the defective part(s).

The materials management organization should have an emergency response system in place for just such contingencies. For example, when the material has been rejected by Q.C. then the bells and whistles should place in motion a reaction, which depends on the lead time for the supply of the items. Such steps might be, as follows:

- Q.C. determines that the material is defective.
- Q.C. places a hold on the material.
- The Q.C. tagged material is moved into a "hold" area of the warehouse or plant.
- Q.C. notifies the buyer.
- Buyer notifies the vendor.

This procedure presupposed that the time lapse from the receipt of the material to notification of the vendor is a matter of hours. The Q.C. staff must inspect the material as soon as possible after its arrival on the receiving dock floor.

The coordination between the Q.C. staff, materials movement, and purchasing must leave nothing to be desired; quick action is extremely important. The speed with which these steps are taken is one way to ensure that the flow of materials to and through the plant is at its optimal momentum. After all, that is the secret of good materials management: the faster the turnover, the greater the profitability of the operation. The same thoroughness of application applies to both materials and equipment.

An example of this type of coordinated approach can best be illustrated by citing an example from my experience following World War II. The Marshall Plan's Lend-Lease Program had supplied a steel foundry in northeastern France with a *sandslinger*, a large piece of foundry machinery used for reproducing the action of a molder in filling a sand casting mold by hand. Sand is delivered in wads at high speed by centrifugal force and directed by the operator into the mold as required. Three months after the sandslinger had been delivered to the plant, it was still in the large crate in which it had been shipped. If the company had prepared the space and the personnel, the sandslinger could have been uncrated and placed into position in a matter of hours upon its arrival at the plant, and the resultant increase in sand-molding productivity would have quadrupled in the three months that had elapsed since the receipt of the machine.

The buyer of materials is no less culpable than the buyer of equipment in taking fast remedial response to a given situation that might cause a lapse in the flow of materials or a decrease in the achievable level of productivity. Thus, if a buyer delays noting the inspection report with its rejection of the material for an unreasonable length of time, he or she is placing the company in an untenable position when it comes to validating the warranty of the seller's merchandise.

Frequently, however, it is possible for the buyer's company to put the materials through a reprocessing operation that may make the material serviceable. In such circumstances, the buyer should come to terms with the vendor prior to the refabricating or repair work and decide how much it would be worth to do so. If the seller agrees that the charges are reasonable, the buyer can order the material repaired; otherwise, the vendor can request that the material be returned to his plant and other means of restitution can be sought. Ordinarily, if the buyer is hard pressed for the material, he or she may buy against the vendor's order. But this is always a last resort, primarily because it leaves a certain amount of residual ill will.

Today's business climate has strengthened the buyer's position in terms of product failure because of the new interpretations of the courts in considering questions of liability. Thus, the buyer is protected, in part, against damages to persons or property that result from failure of the vendor's equipment, as long as there is a reasonable amount of evidence to the effect that failure was due in large part by the seller of the equipment, rather than by the purchaser. This leads to the next part of the purchasing conundrum: the sample.

Samples   In many instances, the vendor will supply the buyer with a sample of the item to be sold, since this is, in effect, putting his or her best foot forward. This fact gives the buyer an implied warranty that the goods to be delivered will be exactly like the sample. The Uniform Commercial Code indicates as much when it states, "any sample or model which is made part of the basis of the bargain (read, *agreement*) creates an implied warranty that the whole of the goods shall conform to the sample or model." Since the sample may become prima facie evidence of the quality of the product to be delivered, it goes without saying that the buyer should carefully catalog and document (with a tag, if necessary) the sample. Many companies, and governmental agencies also, maintain a Sample Room, usually kept under lock and key to make certain that the samples are there when needed for comparison or in matters of litigation.

Specifications   In general, there are two major categories of specifications: functional, and detail. Another category of specification is implied in many off-the-shelf parts, such as fasteners (the nuts and bolts of industry), plumbing fittings, pipe, and electrical components, which are almost completely covered by industry standards that have been prepared by the technical committees of the American National Standards Institute (ANSI), the American Society for Testing and Materials (ASTM), the American Iron and Steel Institute (AISI), and the National Electrical Manufacturers Association (NEMA), to name a few. There is hardly any possibility of legal contention due to poor quality on purchases of such common items. Whenever a batch of such parts is received that does not pass inspection, it is simply returned to the vendor, in most instances with no questions asked.

Functional   Functional specifications, sometimes called *performance specifications*, are used when the achievement of the results or performance of the equipment is known or desired, such as the production of so many pieces per hour or linear feet per minute, and the design, engineering, and fabrication of the machine are entirely the responsibility of the vendor. Although the fabrication of the machine can be left to a third party, the essence of the functional specification is to obtain a device that will accomplish the result specified.

The following text illustrates what is intended in a functional specification. Only a part of the specification or request for proposal is contained in the following paragraphs for the sake of brevity.

This specification covers the minimum requirement for a fastener inspection machine, to be used for dimensional inspection and sorting of titanium fasteners. The manufacturer shall have the choice of designing the machine functions. The machine shall operate automatically after the initial operator setup.

All primary operator control buttons, switches, levers, and indicators shall be functionally grouped and mounted in location providing the operator with the maximum visibility of the machine and tooling from the normal work position.

The machine shall have integral leveling jacks and hold-down bolts to facilitate rapid installation and maintenance of leveling.

The inspected fasteners shall not be deformed or have the anodized finish scratched during the inspection and sorting process.

The fastener inspection machine shall gauge the following fastener parameters as defined by the standard included in Appendix 1 and 2 of this specification:

A—Head diameter
B—Head height
C—Head angle
D—Shank diameter

The inspection machine shall have the following minimum accuracy for the gauge parameters, as follows:

A (head diameter)	±0.001 inch
B (head height)	±0.001 inch
C (head angle)	±0.1 degree
D (shank diameter)	±0.001 inch

The fastener inspection machine shall be capable of inspecting and sorting the following sizes of fasteners, as follows:

Standard Number (size)	Nominal Thread Size (inches)	Nominal Length (in 0.0625-inch increments)
6	.1900 – 32	.75 to 1.625
8	.2500 – 28	.75 to 1.625
10	.3125 – 24	.75 to 1.625

The fastener inspection machine shall be capable of automatically inspecting and sorting a minimum of 30 fasteners per minute.

Sufficient fastener storage shall be provided for a 1/2-hour production run without refilling the feed hopper(s) or emptying the accept/reject bin(s).

The maximum machine setup time, from one standard number (nominal thread size) and nominal length fastener to another shall not exceed 10 minutes and shall be accomplished by the operator without requiring special skills or tools.

There would be much more descriptive verbiage than this in the complete functional description of the fastener inspection machine just described above if you were to actively seek a manufacturer who had the necessary research and development skills to design and build this electromechanical device.

Detailed In a detailed specification, the buyer gives the supplier (manufacturer) a set of blueprints and specifications that describes by specific dimensions and/or specifications and standards. The purchase order (purchase contract) would indicate that the item is to be fabricated exactly in accordance with the specific drawing number(s) and/or specifications, which are made part of the order. With the detailed specifications as a guide, the quality control department need only follow the given specifications in order to accept or reject the product. If the item conforms to specification and the function of the item does not perform satisfactorily, the buyer would be in an extremely awkward position to press his case for performance. The opposite would be true for the buyer who uses a performance (functional) specification.

Other provisions can be included in purchase contracts by both buyers and sellers. For example, either party may desire to include definite provisions in the purchase contract governing warranty limitations. The seller would like to limit his or her liability to the actual cost of the product or item he or she is selling. Since both the buyer and the seller have a sizeable stake in warranty provisions, it is much better to clearly define all the terms involved, even to the point of specifying inspection procedures.

Delivery Of paramount importance to the materials management department is that incoming and outgoing shipments be maintained on a precise schedule. When the materials management team is viewing deliveries from a purchasing standpoint, the due-ins, or receivables, are placed in the spotlight. Naturally, you would expect that the date the material is delivered would be carefully spelled out in the purchase order. Thus, if the vendor accepts the order, then he or she is bound by the contract and promises to deliver by that date. Should the vendor fail to deliver the merchandise by the date prescribed, the laws and regulations of most states indicate that if the seller wrongfully neglects or refuses to deliver the goods, the buyer may take action against the seller for damages resulting from nondelivery.

If the goods that the seller does not deliver are readily available, the buyer may collect general damages, which are limited to the difference between the contracted price and the price that the buyer pays in the marketplace. When the buyer has to buy against the contract, he or she should exercise care to obtain the best possible, competitive price in order to avoid the appearance of high-handedness, simply because the seller will have to make up the difference between the contract price and whatever the buyer must pay for the goods in the marketplace. When it appears that the goods are not readily available, then the buyer may recover special damages. The buyer has to prove that the damages the company has suffered from the failure of the vendor to deliver the goods could reasonably be anticipated by both parties; in addition, the buyer must be able to justify the amount of the damages.

For the material control personnel in the materials management department (including the materials handling personnel), a late delivery can be almost as serious a problem as failure of the vendor to deliver. If the buyer has been slipshod in his or her purchase contract wordage by not indicating a specific delivery date, then by the laws and regulations of most states, the seller may deliver the goods within a *reasonable* time. Since the word *reasonable* lends itself to as many interpretations in the legal process as there are lawyers, it would pay the buyer to guard against late deliveries by including the following phrase (or, words to the same effect) in each purchase contract:

> If shipment is not made at, or before, the time specified, the buyer reserves the right to cancel the order, or any part thereof, without obligation.

If the material control department does not wish to receive the material prior to the specified date, for various reasons, then the words *or before* should be eliminated from this purchase clause. Some companies, as well as most governmental agencies, often require that performance bonds be posted by suppliers prior to contract acceptance, or require that a check for the amount of the bond be submitted along with their response to an invitation to bid. The supplier might have to forfeit all or a part of the bond money if the contract merchandise is not delivered on schedule.

Penalty clauses also can be added to purchase contracts to make late deliveries less profitable for the vendor; the seller's price can be reduced by a specified amount for each day (or, week) that the contractor is behind schedule. The following paragraph is the actual wording of a contract between the city of Tukwila, Washington, as buyer, and a supplier:

> The Contractor shall furnish to the City prior to start of construction a performance bond in an amount of 100% of the contract in a form acceptable to the City. In lieu of bond for contracts less than $25,000, the City may, at the Contractor's option, hold 50% of the contract amount as retainer for a period of 30 days after final acceptance or until receipt of all necessary releases from the Department of Revenue and the Department of Labor and Industries and settlement of any liens, whichever is later.

In this contract clause, there was no mention made of a penalty for late delivery; however, you will notice that 50 percent of the contract amount was to have been retained until final acceptance of the work was given.

Other difficulties   As you know, lawyers become wealthy because of other people's difficulties. There are three gray areas that should not be left without discussion. One subject has to do with discrimination among customers. The materials manager, through his or her purchasing department, should be well aware of the pitfalls. A second area is in the execution of the Fair Labor Standards Act. A third area is the subject of patent rights, which can sometimes be unknowingly violated. A brief summary of each of these areas follows.

Discrimination   The Robinson-Patman Act was designed primarily to prevent a seller from discriminating unfairly among his or her customers. Thus, it is against this law, in most instances, for the seller to offer the same (identical) merchandise in the same quantities to several customers at different prices. On the obverse side of the coin, it is also unlawful for the buyer to participate in such discrimination, where a competitive advantage is gained in this manner.

It should not be construed that buyers cannot shop around for better—i.e., lower—prices; however, care must be taken to avoid duplicity in the matter.

It is one of the incontrovertible facts of purchasing that some buyers will play one seller against another to obtain lower prices, usually on small-item purchases and in telephone bids. Although most purchasing agents do not condone this behavior, it is done every day. This practice apparently does not violate the act, however unethical price undercutting may sound. Also, if the buyer is able to purchase items in larger quantities that will enable the supplier to achieve lower production costs, then he or she does not violate the spirit of the law. The chief benefit of the act has been to guard against monopolistic pricing policies of some small number of unscrupulous suppliers and to keep the majority of suppliers honest.

Fair Labor Standards Act   The Fair Labor Standards Act has been effective in regulating child labor, wages, and hours. Every state in the union has this type of legislation spelled out, and with the increase in emphasis on the labor practices involving women and minorities, almost every formal purchase order printed bears some indication of the protective certification required of the sellers. Where standard boilerplated forms, which are available in most states, are not used, company attorneys should be consulted in order to obtain the latest wording required by law to protect the buyer from possible litigation.

Patent rights   The U.S. government patent law has given an inventor a legal right to manufacture or assign all or part of that right to another for a period of 17 years. Some patents may be improved by the inventor after this period of time for additional years of protection. In addition, the U.S. government has entered into reciprocal arrangements with some foreign countries that may extend the inventor's rights into these countries also. Thus, if any other person or company manufactures, sells, or uses the patented item during the 17-year period, they are guilty of patent infringement, and the inventor may sue them and collect damages.

Purchasers may unknowingly buy an item manufactured by a supplier who is infringing upon the inventor's patent and thus be liable for damages, as well. Although this occurrence is relatively infrequent, some buyers have wished to avoid even the rare occasion for error in both copyrights and patent rights and have used the following clause in some purchasing contracts:

> Seller warrants that there has been no violation of copyrights or patent rights in manufacturing, producing, or selling the items shipped or ordered, and seller agrees to hold harmless the purchaser from any and all liability, loss, or expense occasioned by any such violation.

As you can see, the patent rights might cover the means or method of producing the item. The insertion of this clause in the purchasing contract will not necessarily enable the buyer to transfer liability to the supplier who has infringed upon the inventor's patent, but it can give the buyer a basis for a claim against the supplier if the inventor sues him or her for damages.

The practice of patent law is a highly specialized field of legal activity and, as such, deserves special attention by the purchasing department. Since it is so specialized, the company's attorney should be consulted whenever there is any doubt, or when a problem arises that falls in this area of concern.

Tax exemption   A federal excise tax is imposed on some goods that are used by the company. The exceptions to this form of taxation are those items that enter into the production process or form part of an end product. If a buyer wants to avoid the tax (which is a necessary cost avoidance), the buyer must include a tax exemption certificate with the purchase order. If a company is using purchased materials directly for resale—i.e., without using them in the manufacturing process—it is also exempt from the federal excise tax and should add an exemption clause to its purchase order. (A special exemption clause would have to be included if the material in question is to be exported.)

Since almost all states have sales or use taxes, if material is purchased for resale or for incorporation into another product, it might be exempt from the tax. But since the state laws vary somewhat from state to state, a different clause might be required for any state in which sales are made or the buyer does business. The company attorney should be relied upon for this information.

In FIG. 4-6 which illustrates a typical, large company purchase order, the sales tax status is indicated as being applied for the state of Washington. A further clause indicates that:

> Any federal, state, local, or excise taxes for articles or services obtained by this purchase order must be identified as a separate item on each invoice submitted.

Purchasing and the materials manager

In chapter 1, the various forms of organizational structure were discussed in order to obtain a perspective of materials management. This chapter has been developed to flesh out the thesis that some one person in the company's organization should be responsible for making decisions concerning the acquisition and flow of materials from outside suppliers to and through the company's manufacturing processing, storage, and distribution network.

Since all of these functions are interrelated, it appears to be the logical conclusion that resting authority for materials movement and, ergo, management, upon one person will eliminate many problems that have to do with overlapping functions and pride of turf where each of the entities has responsibilities in its own domain and desires to maintain the lowest possible costs for its department. Since the production control and inventory managers want to perform at their best, both of these departments do not want to run out of materials. Material shortages and operating on a "short sheet" mean that labor effort must be wasted on expediting, wherein the material control inventory personnel make frantic phone calls and visits to various departments and harass the purchasing personnel to call suppliers. This situation is where pricing and costs soar out of control.

Since the production control and inventory control managers do not have the responsibility for negotiating prices, they must leave this task to the purchasing department. The purchaser (buyer) negotiates prices based on quantities that are determined by the material control department, and any shortfalls that require emergency (expedited) buying are the bane of the purchasing department's existence. If the buyer purchases materials from a supplier in an area that is poorly served by either rail or motor freight, the traffic department might not be able to obtain favorable freight rates.

The logic of having the flow of materials organized so that materials control, materials handling, production control, inventory, purchasing, and traffic departments are all under the single authority of the materials manager becomes difficult to refute, especially when it can so definitively rule out overlapping functions with the resultant turf wars that sometimes take place, even in relatively well managed companies. The materials manager who has the purchasing department as part of his or her organization is a fortunate individual because he or she then has the clout and the corporate position to make important contributions to the economical operation of the company. Also, when all the departments involved in material flow are under one roof, the buck-passing that can be self-destructive to a company is minimized, if not eliminated completely.

In some companies, the person who heads up the purchasing department becomes the materials manager ipso facto because through a process of his or her expanding authority or by acquiring added responsibilities, it sometimes develops that not only purchasing, but also traffic and inventory control are placed under his or her control. In essence, the purchasing director is well on his or her way to becoming a full-fledged materials manager.

To complete the materials movement package, it is necessary to add the responsibility for all the internal plant handling that revolves around materials control, such that materials handling, production control, and distribution (sometimes) are additional functions of the materials manager, who might have started his or her enclave by being a top manager, usually a vice-president, in charge of purchasing and traffic. Nevertheless, with the status of the position, materials management becomes a no-longer-muted voice in the top echelons of management, and the materials flow process can be conducted as an integrated systems network with a chain of command that goes directly to the top of the organization.

To become an effective materials manager, however, a purchasing director needs expertise in areas that are not usually associated with purchasing. For example, it is an indisputable fact that production and inventory control cannot be conveniently dissociated from manufacturing. In the case of inventory control, purchasing has a large impact on this function simply because the quantities of materials purchased determine the levels of inventory. Also, the levels of inventory have an effect on production (manufacturing) schedules; thus, the production control department function is closely related, in this manner, to both inventory control and purchasing because of in-process (work-in-process), semi-finished, and raw materials inventories.

Materials flow functions such as materials handling, receiving, shipping, warehousing, and distribution do not ordinarily command much attention from top management when they are part of the manufacturing organization unless they become trouble spots. Therefore, without a materials management organization, these departments are subordinated to the principal objective of manufacturing, which is making the product. The professional materials manager is more than a purchasing director in terms of the breadth of understanding that is associated with, and required of, the function. For this reason, when material flow throughout the company is headed by a materials manager, morale in all departments tends to improve, proving the corollary that the whole is greater than the sum of all its parts.

5

Engineering and design as it affects materials management

Introduction

As described earlier in the text, materials management has as its overall objective the orderly flow of materials from the source of supply, through manufacturing, to the ultimate consumer. The materials manager has his or her hand on the faucet knob that regulates this flow in proportion to the demands of the marketplace and the availability of materials. He or she is influenced in this regulatory function by such things as the price and quality of materials, delivery schedules, and the internal concerns of the company. He or she has to make decisions based upon the data that is fed to the materials management organization by a number of other departments of the company, as well as by suppliers and other informed sources.

The materials manager has to be aware of business news and the condition of the markets, just as the marketing department does. If the marketing department does its task well, then the materials manager has a depth of information to draw upon; in any case, the materials manager must be informed and knowledgeable concerning the vagaries of the marketplace.

In a simplistic summary, the materials manager has to keep relatively well informed of both the internal status of the company and the economic forces outside the company that, in the final analysis, control the destiny of the organization. He or she must decide how much material to obtain, when to obtain it, and from whom to obtain it. When juxtaposed against all the variables involved in the process, the task is no longer as simple as it first appears.

Designing the product

Even the simplest products require many technical skills in both the design and engineering phases, as well as in the fabrication aspects. The common electric toaster, for example, represents a remarkable variety of inputs from materials to processes. Consider for a minute the amount of nickel-chromium alloys, or nichrome wire, which is the electrical resistance designed by someone with electrical talent; the ceramic dielectrics that the wire is threaded through; the springs and thermistors; the metal stampings forming the shell of the device; the decorative and rust-inhibiting chrome plating; and the plastic parts for knobs and dials and the like. Packaging materials to house the completed machine would be comprised of an inner pack and an outer pack. The outer pack is the shipping container necessary to get the finished product to the wholesaler's warehouse or outlet store.

But this is only the tip of the iceberg because machinery is required to produce the appliance and spare parts for the machinery. Service contracts to maintain it must be obtained unless maintenance is an in-house operation. Plating tanks and supplies must be purchased. Steam boilers, generators, and provision for storing hazardous materials must be made, in addition to arrangements for the safe disposal of the effluents and by-products of the plating processes.

Materials management, in even some of the smaller companies, can involve many hundreds and thousands of parts and materials, some purchased raw, some semifinished, and some finished. The principal reason for the pyramiding of the numbers for raw materials, for parts and components, and for equipment—since different end-products might require several types of fabricating machinery—is that an appliance manufacturer, for example, would make not only an electric toaster, but also a whole family of consumer goods from toasters to mixing machines, to blenders, and so forth.

In most companies the major materials management effort is devoted to materials that are used directly in the products themselves, hence the term *direct materials*. As much as three-quarters of the purchasing dollar is spent on direct materials, and in general, according to Pareto's Law, about 20 percent of the items result in 80 percent of the expenditures of the company. Thus, the purchasing department will monitor the 20 percent portion of the items vigorously, leaving the purchasing of these items to the most senior buyers. Although purchasing works closely with inventory control on all items, special emphasis is given to the high-dollar-value items.

In well-organized materials management departments, a materials manager will have delegated authority to the extent that minor problems are recognized as such and decisions affecting these difficulties are handled by subordinates on a relatively low level in the chain of command. Major problems, on the other hand, have to be recognized early on, and as such should be directed to the materials manager.

The exception to this rule occurs in the early design stages of a product. At the beginning of the design period, decisions on materials and processes can have a monumental effect on every stage thereafter. Because of the importance of design decisions made at this point, questions concerning sourcing, for instance, can limit the flexibility of suppliers that are available to the materials department for long into the future. In addition, improper design decisions can create numerous problems that are difficult to rectify satisfactorily. For this reason, the materials manager should keep in close touch with the engineering designers who are launching any new product or product line.

In large companies where there is a research group, or research and development (R&D) organization, once the feasibility of a product has been determined, it is up to the engineering department to "put it into iron." Smaller companies perform their R&D and engineering design in the same unit, but the results are usually the same, except for a time lapse. Some small companies can proceed from the concept stage to the prototype in a matter of months. Usually, in larger companies, anywhere from two to five years to bring a product to market would be considered more realistic.

Despite the type of R&D and engineering organization a company maintains, almost every company depends to a large extent on its cadre of suppliers for some portion of the design effort. This is where the materials manager and his or her department bring their expertise to the forefront because, no matter how large the company in question, it probably depends on its suppliers to produce some of the parts and components that are to be used—i.e., assembled—into the end product.

As mentioned previously, suppliers tend to play a much larger part in new product development than the general public realizes. No company is large enough, nor is the government itself capable of assembling all the scientists and engineers that might contribute to the end product. When I first visited the Huntsville, Alabama, Space Flight Center prior to the final assembly of the initial Saturn rocket, over 100 major contractors were employed on the project. Each of these majors had numerous subcontractors who were producing parts and components for them. Possibly the only exceptions to this general observation are service industries and some not-for-profit—i.e., eleemosynary—institutions.

The automobile industry is an excellent example of an industry's dependence on a strong supplier network. Many automobile components were developed (and continue being developed) by suppliers—from the common auto tire, to transmissions, power steering, wheels and wheel covers, turbochargers, generators, alternators, and electrical components. I worked on patents for one of the earlier turn-signal devices that are commonplace today. Devices and parts now fabricated for the automotive industry are the mainstay of whole cities. The fortunes of Cleveland, Ohio, rise and fall with the fluctuations of the auto industry, as an example.

Thus, the materials management organization, through its materials control and purchasing departments, should provide the necessary liaison among knowledgeable and proficient suppliers; the designers, engineers, and technicians in the company's engineering department; and the R&D effort. It is a waste of the company's time and money to reinvent the wheel, and bringing supplier know-how to bear on the company's design and technical problems surely will pay off in the long run.

Styling

Generally speaking, the matter of styling has gained more emphasis as time goes on. You need only look at the way automobiles and other consumer durables have developed over the years, not to mention all the small appliances and containers for products that are mass-produced all over the world. The biggest impact in the company's organization is the influence exerted by the marketing group, which really sets the tone and style of the product, from forklift and off-highway trucks to perfume bottles.

Our electric toaster, in a prior example, is as good an example as any for the fundamental thesis of styling. A graphics design artist makes a number of sketches of the proposed toaster from inputs derived from marketing studies. A company review of the subject where engineering, marketing, stylists (designers), and every concerned department of the company is represented must approve and finalize the design. At this "review" meeting, prior work has been done to include inputs from materials management to determine sourcing and materials substitutions, since the final design(s) to be presented must have the maximum sales appeal consistent with the lowest possible cost. The margin of profitability must be kept as high as possible without sacrificing quality.

By analyzing comparable designs it is possible to determine the probable cost of each model in a satisfactory manner. Although these costs are relative measures, they are generally within 5 percent or less of the true cost. It is only with relatively complex machinery and devices that cost overruns might sometimes become burdensome. Much of the time, the cost for new products can be determined fairly closely by comparing the differences and complexities of the new design with similar or older models.

Some of the factors influencing costs, of course, are the differences in materials that are used to make the product. This is where the experience of purchasing personnel, under the aegis of the materials manager, can make a valuable contribution to the design effort because different styling requirements based upon materials can be estimated by the materials buyers with a fair degree of precision since they are constantly comparing values in their daily activities.

In evaluating various styling efforts, it is assumed that several competing materials can be used, as in the following list:

- High-impact polyethylene
- Zinc die casting
- Sheet steel stamping
- Aluminum die casting

If the product can be made satisfactorily from any of these materials, the decision is relegated from a strictly engineering value decision to one that is purely economic.

Another advantage, when several designs are reviewed by experienced materials personnel, is that since the relative costs of several materials can be compared it sometimes happens that through value engineering studies and the like, minor changes in the design can be suggested that will lower the cost of fabrication. A shallow-draw die, rather than a deep-draw die, might cut manufacturing time in half for the given part, or other facets of the shape of the part might be suggested to lower costs. In preliminary reviews of this nature, it might be possible to eliminate many costly design features that do not enhance either the styling or the functionality of the part.

Since styling usually consists of the outer layer of the device—the skin, the framework, the monocoque body of an automobile, etc.—it sometimes dictates the size and function of the mechanism that is hidden. In a forktruck, for instance, the overhead guard, the operator's compartment, and sometimes the counterweight have been shaped by the stylist. It then becomes the engineer's task to specify the artist's view of the machine into something that can be fabricated readily. As more than one builder-contractor has stated of architects' plans, just because it's on the blueprint doesn't mean it can be built.

For this reason, the material manager and his or her staff should closely monitor the earliest stages of the development of any product so they can guide the stylists and designers by means of cost-effective suggestions; by the use of standard, off-the-shelf components; and by use of suppliers to assist in offering their company's expertise and products that can be integrated into the new product. This is a very productive use of the purchasing staff's expertise, and it helps immensely in guiding design decisions and exposing the company's engineers and technicians to new products, new materials, and new processes.

Standard parts

Almost every industry has standard components that are produced in such large quantities that the cost of each unit is relatively low. Good examples are automobile batteries, tires, and spark plugs. In the electronics field there are hundreds and even thousands of standard parts, such as programmable controllers, resistors, and diodes. Therefore, wherever a company can avail itself of standard componetry in a new product, it is well advised to do so. The purchasing group, who is close to the marketplace in terms of what is and is not standard (for example size, shape, horsepower, and voltage), can be invaluable when a new product is being reviewed from this standpoint.

When company technicians obtain developmental assistance from a supplier, in effect using the specialized know-how and expertise of the supplier, it is very easy to become locked into that particular supplier since he or she is usually desirous of incorporating his or her company specifications into the new product. If this happens then competitive bidding on components for the new product might be severely restricted.

This is one of the main reasons for the materials management department's overview role in the development process. If both the engineering department and materials management are doing their jobs well, it will be possible to develop a strong supplier base to ensure competitive pricing on all the materials and components of the new product.

Another aspect of the situation is the possibility that the engineering department will have alternative materials and component specifications to choose from in assembling the new product. When this fortunate situation arises, which is not uncommon in design work, the materials management group can make important contributions from its study of the availability of materials, material and component market prices, and the like, and the materials managers can advise the engineers accordingly.

Also, a word about market prices: if the materials management group is doing its very best in terms of analyzing market trends it can sometimes advise against a particular lower-priced material in the current market based on a long-term trend that might see a somewhat currently higher-priced material having a cost advantage in the future. This factor was especially true in the plastics marketplace just a short time ago; however, in the light of rising world petroleum prices, some petrochemical materials might be priced out of the market in the long haul.

Materials and processes

Every engineer tries to keep current on new materials and processes. One of the perplexing problems in the engineering community today is that technology is advancing so rapidly on all fronts that it is virtually impossible for even the most knowledgeable technicians to keep up to date on even the narrow focus of their particular specialty, given the worldwide pursuit of new products, materials, and processes. Since many different industries and companies are concerned in these endeavors, even some computer networks fail to contain everything the engineer needs to know about his or her field of specialization.

This is another area where the materials management department can serve the engineering organization, by assisting in updating the technicians on the latest developments from suppliers. Often, slight changes in materials or processes can improve the quality of the company's end product. Here again, liaison performed by the materials manager between the engineering and quality control departments will pay enormous dividends for the company in terms of enhancing the quality and, therefore, the sales appeal of the item produce.

Make-or-buy decisions

In chapter 3, there was a brief discussion of the "make-or-buy" philosophy as it touched on the interrelationships between manufacturing and materials management. To make or to buy a part, component, or product is never a unilateral decision by one department. Far from it, primarily because it might increase or decrease costs to the extent that profitability might be enhanced or decreased according to the way the economic weather vane turns.

Another aspect of the make-or-buy decision, which is often overlooked by some companies, is that it must be reviewed constantly—it is not a one-time thing. Changing market conditions dictate the manner in which this decision's profitability might be made more or less acceptable. The materials management task in this matter is very clear. It must be firmly imbedded in the department's charter that the decision whether to make the part (or whatever) in-house or to buy the part must be reviewed on a definite periodicity based upon the dollar value of the part. That is to say that a high-dollar-value part must be reviewed frequently, either once a month or at least quarterly, and a low-dollar-value part should be reviewed at least annually, for as long as the part remains in the product domain of the company.

Sometimes spare parts availability might, as a company policy, be in the company's inventory for decades after the product (or model) is no longer in production. In addition, marketing practice might dictate that the after market for the part might be more profitable than the part in the originally issued product.

In my experience, some products with high wearing parts could be sold at cost in order to benefit by the spare parts aftermarket. A product with a 10 percent market price markup could easily have a 200 percent parts markup in the aftermarket. Therein lies a significant contribution to the overall profitability of the company. This type of aftermarket sale is widely practiced in some consumer goods areas, such as household appliances, and is especially true of the automotive and heavy-industry segments of the marketplace.

It is unfortunate that some make-or-buy decisions must take into account the type of equipment that the part requires for manufacture. For example, if a part involves the purchase of new machinery and equipment to produce, and is relatively capital-intensive, the company has some hard choices to make. On the other hand, if the part can be produced on presently available equipment, production of the part might be dependent on machine loading capacity—again, a hard choice to make, depending on present and future production schedules. If the company decides that making the part will place it in a better competitive position or that it might become an industry source for the part, then it might embark wholeheartedly into committing capital for the enterprise.

Some further drawbacks might inhibit commitment of capital for a "make" decision: the lack of in-house expertise on producing the part, or the high cost of capital that would be required. The materials manager's input in the make-or-buy decision becomes a very crucial part of the resolution to the question. If the materials management decision favors the "make," then on a capital-intensive product or part, the company might be committed for years to come because estimates of the availability of materials and present and future prices depend almost entirely upon the materials manager's judgmental abilities. The material manager's degree of exposure in these instances—i.e., when confronted with the task of representing his or her department in the top management review of the make-or-buy decision—requires a certain degree of fortitude.

The difference between the purchased cost of the part and the company's manufacturing costs if it made the part would be the company's *return on investment* (R.O.I.). If, because of competitive market conditions or a decline in the economy as a whole, the purchased price might decline, then the company's R.O.I. would also decline if it chose to fabricate the part. In an upswinging market, prices would rise and, ergo, the rate of return would increase. Thus it is, that when poor or hasty "make" decisions are made, the company's funds are tied up in equipment that yields a poor rate of return and is a lodestone that drains off corporate earnings.

The materials manager has other responsibilities whenever the company management decides that a part, or family of parts, should be manufactured in-house. Most plants do not provide sufficient space within the buildings for new equipment and processes, unless a building has been designed with expansion in mind. The feeling of most managements is that, if expansion is required, they will add on or house the equipment in new or rented facilities.

In some "make" decisions, a facility or building will have to be purchased. The materials manager, together with the other upper management personnel, will have the responsibility of selecting the site. If a new building is to be erected, then the materials manager will negotiate the contract for its construction and serve as the intermediary between the company and the contractors and subs for its erection.

In designing the building, planning engineers in the manufacturing department would largely be responsible for the facility area—the placement of machinery and support equipment. The materials-handling contingent of the materials management organization, however, should work closely with the process and manufacturing engineers to ensure that materials-handling equipment and methods are introduced and integrated with *systems approach* before the first part is made in the new facility.

Integrating materials handling with manufacturing processes is the optimum methodology for ensuring the least total cost of operating the new facility. Thus, the purchasing department contract specialists, working closely with the plant materials-handling group and the process engineers, would develop the specifications for both manufacturing and supportive equipment. The purchasing group would handle the negotiations with the suppliers and administrate the purchase contracts.

As a result of the various complexities involved in make-or-buy decisions, it is apparent that the materials manager had best surround himself or herself with the most capable professionals in their respective fields. In this way, the materials manager's exposure will produce beneficial results and added respect for the position he or she occupies as a top executive of the company.

Simulation

In order to increase the probability of profitability and confirm that the management decision to make the part is the proper one, all of the areas mentioned in the previous section would have to be thoroughly explored. As an adjunct to the close cooperation among purchasing, materials handling, and manufacturing that has resulted in the selection of machinery and supportive equipment, the manufacturing process engineers in consort with the materials-handling group have now turned their efforts to simulating the new facility.

The plant layout is converted to machine coordinates and laid out in computer graphics. The machine speeds and feeds and the parameters involving handling times are fed into the computer after it has been programmed to accept this information. The computer is now capable of taking whatever data and variables (in terms of quantities) that have been described in order to illustrate the path and elemental times involved in the processing.

It is possible to simulate the part, parts, and components as they proceed from machine to machine, from process to process, and watch a progression of parts through the manufacturing process. The process engineers verify their worksheet data from this display, and any aberrations or hangups, where unwanted queues are formed, can be visualized from the computer display.

As a result of the simulation, changes in processing times can be made before the equipment specifications is finalized. Thus, equipment specifications, capacities, and the like can be firmed prior to purchasing, and minimum levels of performance can be established. Also, changes in processing methods can be made prior to the initial product runs. Other changes in plant and equipment also can be made at this time, and certainly made more easily than they ever could after the equipment has been purchased.

6

Requirements planning methodologies

Introduction

A recent advertisement in one of the leading manufacturing systems magazines makes an offer to its readers to start them off in materials requirements planning (MRP) for only $195. The ad goes on to explain that they (the company, called Micro-MRP, Inc.[1]) have one of the best-selling Foundation Systems, composed of:

- Bill of Materials
- Inventory Control
- Interface

The offer is, of course, for a packaged computer program. Its effectiveness is said to be such that it is possible for your company to save enough money in a year's time, or less, to be enabled to purchase "Micro-MAX MRP™," which is the complete MRPII planning, production, and control system. (MRPII is the abbreviation for Material Resource Planning.)

The chronology of development for these systems is as follows:

- MRP (or MRPI to differentiate it from MRPII), an abbreviation for Material Requirements Planning.
- MRPII, an abbreviation for Material Resource Planning.

There is also Closed-Loop MRP, Distribution Resource Planning (DRP), and Capacity Requirements Planning (CRP). Each of these systems satisfies a particular requirement in the art of planning; however, by far the most important and widely used technique is the MRPII concept. As the ad states, you can get started using MRP methodology using an IBM microcomputer (PC, XT, AT, 386, or any 100 percent IBM-compatible micro) equipped with 640K memory and hard-disk storage.

In this particular MRP program you are able to maintain a manufacturing database and a completely formalized parts list, which fully describes the product, defines component relationships, and determines where each part is used. With this program you also can develop standard costs and improved product pricing. The bill of materials segment enables you to identify obsolete parts so that they can be swiftly eliminated from stock. In addition, you can track your inventory and get a full, part-usage history, by part, by job, or by user-defined stock location. It enables you to uncover inventory discrepancies, and will automatically signal reorder points, thereby increasing clerical productivity. The interface provided in the program permits you to analyze data on dBASE®, Lotus 1 – 2 – 3®, and other popular software, as well as create graphics and customized reports in your own business formats. In addition, you can transfer data to and from other computers and systems (from a mainframe at corporate headquarters, for example, to a PC at the department level). You can download frequently referenced data to your IBM micro in order to decrease file maintenance and transmission costs, and further tighten manufacturing control.

MRPI, MRPII, and DRP are primarily planning systems that were developed for use in solving the complex logistics problems arising in the manufacturing environment. All of these systems require current, up-to-the-minute (real-time) data, which is collected by shop floor automation networks in order to be functional and useful. This requirement has led to the disposition of computer terminals throughout the manufacturing area, the recognition that people who do the planning must have timely data from the factory, and the understanding that factory people—i.e., those in manufacturing on the shop floor—must depend on planners in order to completely integrate the two domains into a satisfactory whole.

Material requirements planning (MRPI)

Until fairly recently, MRP has been viewed as, more or less, "exploding" the bill of materials in order to produce valid manufacturing schedules. Currently, it is seen as a device for management to control cash flow, to decrease and trim the fat from inventories. It is also helpful in making material purchases, and is instrumental in using the work force more productively than was possible a decade or so ago.

In the earlier, simplistic view, it was used primarily as an ordering methodology. The rapid developments in computer technology were a major influence in bringing it into prominence.

Despite its utility, however, MRP planning and concomitant programs are virtually nonexistent in smaller companies. Even in some of the larger and multinational companies there doesn't appear to be a full-scale embracement of the methodology, but what seems to be partial implementation for the most part.

A certain amount of validity can be ascribed to this apparent partial adoption of the MRP method, for the simple reason that it requires a vast investment in company time, endeavor, and complete involvement of at least a portion of the company's personnel for it to become a companywide program. Once the company decides to mount the MRP bandwagon, it is necessary to enroll all of the company into the MRP philosophy because the Master Production Schedule, which is the motor of the MRP system, is the embodiment of the total requirements of the company based on the sales forecast, actual demand, master construction schedule, inventory, and in-house capacity to produce (or build).

No matter how large or small a company is, materials planning is a function that combines production and inventory control. The prime objective of materials planning is to manage and control inventories of parts, components, semi-finished and finished parts, and raw materials that are to be used in the manufacture of products made to order or for stock. It is the overall task of planning, production, and inventory control to establish the quantity of materials required to meet and satisfy manufacturing schedules, and it doesn't make too much difference whether the materials are needed for stock or order, except when deadlines are approaching and the materials are not on hand in the assembly areas, the final assembly line, or the area where parts and components are fabricated and stored.

One of the problems with controlling inventories is that most parts and components are fabricated in batches, usually in relatively small lot sizes. Furthermore, the demand for these materials will occur not continuously, but in intervals, for the most part. To satisfy the demand for the required quantities of materials, it is possible to calculate—based on a projection of sales figures supplied by the marketing group, and then working backwards, exploding the bill of materials for all of the finished product—in neat little groups the number of parts and components required at various intervals of time. Thus, material requirements planning has become essentially an inventory control method specifically designed to meet targeted demand. In addition, the use of computers in manufacturing has enabled the large quantities of data that are generated to be readily handled. It is for this reason that the use of the MRP concept is fairly widespread in industry practice.

The major difference between MRP and the conventional statistical ordering point methodologies used to control inventories lies largely in the fact that MRP is product, rather than part, oriented. Also, it is based on a projection of the demand for the finished product, rather than a review of the historical behavior of its component parts.

The MRP concept begins with forecasts of demand for the finished products and generates time-phased material requirements based on the bill of materials for the products. Since MRP looks at the behavior (or demand for) the whole product, rather than the erratic or spaced movement of individual parts, it is better able to predict requirements and to deal with changes in product demand.

The selection of inventory-control techniques is sometimes based on the kinds of items they manage. For example, a part is considered "independent" if it is not used in a subcomponent, component, or finished product. An example of an independent part would be an item sold as a spare part. When an item or subassembly is part of a finished product, it can be said to be a dependent part when it enters into the calculation of requirements for planning purposes. The MRP concept for inventory control is, in general, based on parts having dependent demand. However, the debate over terminology is academic since provision in manufacturing schedules must be made for both kinds of parts.

Nevertheless, the selection of inventory-control methods is often based on the way in which demand characteristics for the item(s) are perceived. In the main, the statistical order point techniques assume a relatively uniform demand; however, this does not materially affect the number of items in inventory, since a relatively restrictive view makes it necessary to order the materials for replenishment when a certain minimum level is reached. If demand should suddenly slacken after a purchase or production run of the item, the company is left with a relatively large quantity of the items in stock (if the item happened to be an especially fast-moving part prior to the falloff in volume).

It is a fact of life, also, that parts and components used in manufacturing are most economically produced in quantity, the larger the quantity the lower the unit price, all other things being equal. When the intermittent demand for a part is coupled with economical lot size, resorting to the MRP approach—which has the capability of delaying replenishment until the need for the material actually arises—the planners can smooth out production curves and keep inventories at minimal levels without the sharp rises and falls in quantities that generally occur with statistical ordering points, especially when a low level of inventory is reached. Also, MRP planners can better handle the spikes in requirements when similar parts are required simultaneously for several products.

MRP manages inventories of parts by obtaining data that indicate the end product's required parts, the quantity of these parts that are to be obtained (fabricated, purchased, etc.), the sequence in which the parts are to be used or combined in assemblies, and the rate or time in which they are to be consumed. In the overall view of MRP, future demand for all end products based on marketing's forecasted demand is the motor that drives the concept.

Using MRP to control production and inventory

The primary purpose of MRP is for control; the secondary purpose is to ensure the proper flow of materials throughout the manufacturing system. The tertiary effect of MRP is to support planning and execution of the master manufacturing schedule by processing information and generating reports. These three facets of MRP make it a tool of inestimable value for the materials management department because it deals with the flow of materials through the manufacturing process, ensuring that both raw materials and manufactured parts are available when required and in the proper quantities.

In computerizing the MRP concept, it is possible to translate the methodology, which involves a considerable number of feedback loops in order to maintain order in the system, into a graphic display, as shown in FIG. 6-1.

As mentioned earlier, shop floor terminals are the important links in the system; however, terminals must be located strategically to impart the necessary data required in the MRP database. The master schedule, being the plan for production, is pivotal in its position in the network. Although you cannot say one thing is more important than another in a given system, inventory records require a remarkable degree of precision, since the balances of materials, parts, and components on hand or on order are what keep a well-run plant on target.

Since lead times for the receipt of materials from vendors must be factored into every plan, what happens to inventories in the interim between notifying the purchasing department (through the requisitioning process) to the final receipt of materials depends on how well the inventory control group really controls inventories. Lead times

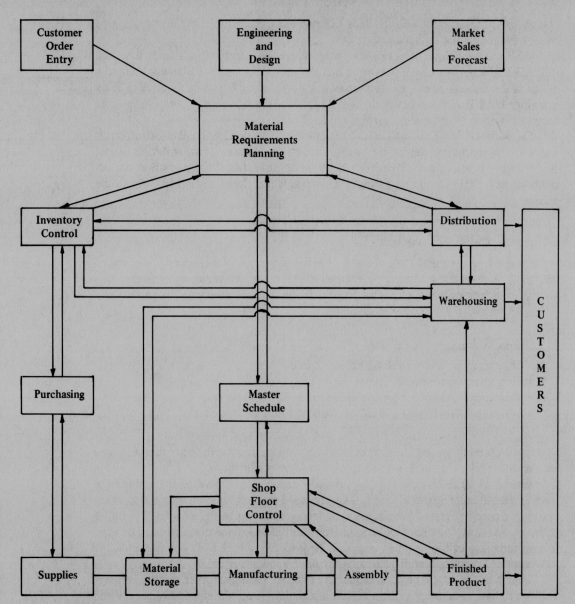

Fig. 6-1. Feedback patterns in material requirements planning.

can really play havoc with manufacturing schedules when vendors fail to perform reliably.

Thus, it is easy to see how important the purchasing function is to MRP and materials management, since it is one of the primary tasks of purchasing to qualify suppliers and evaluate their performance. As a chain is no stronger than its weakest link, so is the dependency of each facet of the materials management organization on all members of the group. If these close relationships did not exist, then the company would save money by eliminating redundancy.

MRP can be thought of as the brain that makes the manufacturing body function. The MRP planners receive inputs from many sources, the primary element being the bill of materials. As a result, it is important that engineering inputs be received to bring drawings and bills of materials up to date. Engineering design change notices should be received as soon as approved drawings are finalized.

The output of MRP consists of instructions to put the production plan into effect. It provides the purchasing department with schedules for obtaining raw materials, parts, and the like. It also gives instructions to the shop indicating what parts and components must be fabricated and assembled, in what quantities, and within what time frame.

Phases in the MRP process

Phase I   At the first step, phase I, of MRP, the normal functioning of the manufacturing process is simulated by computer. Through the simulation of the flow of materials through the plant—i.e., through all of its manufacturing processes—MRP determines the following elements:

- What materials are required
- How many pieces of each part are required
- How much material is required
- How many component groups are required
- When these materials, pieces, components, etc., are required

MRP uses the forecast of the required demand for the finished product and then explodes the bills of materials of each product to be manufactured.

Phase II   The objective of the manufacturing department is to produce the proper quantities of quality products—i.e., finished product and items for stock, spare parts, and components—if the plant is a made-to-stock company, as illustrated in FIG. 6-2. In *made-to-stock*, the total production of the plant is based, for the most part, on marketing forecasts of the volumes required. A *made-to-order* plant is one in which the company will produce a product based only upon a customer order.

That is not to say that a certain quantity of spare parts and other components will not be produced according to arbitrary ratios that have been developed historically and fine-tuned over the course of years of experience in the industry. Also, some speculative units of product might be added to the plant's build schedule, but in the main, a made-to-order plant is relatively conservative regarding product volume. Big ticket items—i.e., those in which the price of each unit is relatively large—tend to make manufacturers fairly cautious in this regard.

This does not necessarily inhibit large, multinational companies, who have sufficient capital resources and a virtually unlimited line of credit, from entering the marketplace with speculative quantities of product, since any number of these companies will floor-plan, or encourage (sometimes coerce) their dealers into taking a number of

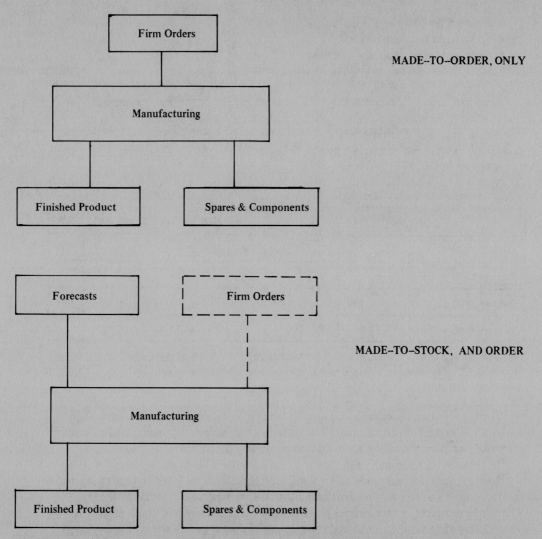

Fig. 6-2. Two kinds of manufacturing philosophies: made-to-order only, and made-to-stock-and-order.

high-price units on consignment with no payments required until the units are sold or sometime require them to purchase the units after a specified period of time has elapsed.

For the most part, companies manufacture product for stock; however, the MRP procedure remains virtually the same regardless of the operating philosophy. Therefore, a time-phased master schedule is prepared so that the desired quantities of finished stock can be produced. This master schedule usually has the following components:

- Demand forecasts
- Customer orders

- Finished product stock requirements
- Spare parts requirements
- Parts and components for inventory
- Stock orders for smoothing production curves.[2]

Since the master schedule is time-phased, it shows the quantity of required production in each fixed interval of time; sometimes the length of the period can be a day, or week(s), or month(s). The time taken up by the complete scheduling period is called the *planning horizon*. An example of a master schedule for a fictitious XYZ Manufacturing Co. is shown in FIG. 6-3.

	AUG	SEPT	OCT	NOV	DEC	JAN	FEB	MAR	APR	MAY	JUN
Product 1000		50				100			100		
Product 2000	25			25			25			25	
Product 3000			100		100			100			100
Product 4000	50			50		25		25		25	

Fig. 6-3. An example of a master schedule showing the time-phase elements and product quantities. (NOTE: The plant was shut down for the summer vacation in July.)

The bill of materials for each "finished product"[3] is the single, most important input to the MRP system. There is one bill of materials for each end product, and it is the summation of all of the bills of materials that comprise the master schedule.

As indicated in the abbreviated master schedule of the XYZ Manufacturing Co., (FIG. 6-3) there are four finished products involved, which must be exploded so that each part can be tracked through the manufacturing process, or purchased, as the case may be. The purchased parts or components do not require bills of materials, since in the MRP methodology they are not considered components. Taken together, the bills of materials (for all of the finished products) show all the components, both purchased and manufactured, at every level of the manufacturing process that are required for the finished product, in addition to the required quantities of each part and component.

An example of a single-level bill is shown in FIG. 6-4. The cart assembly, Drwg. #1001, indicates just one material conversion of the manufacturing process and one material flow phase of the plant.

The compilation of the bill of materials for Product 1000, Mobile Table, clearly that various part numbers require different quantities of materials, as shown in FIG. 6-5. Since the product, in this case a mobile table, is composed of some purchased parts and materials that are to be fabricated in-plant, the bill of materials is multileveled. The diagram shown in FIG. 6-6 indicates in a product structure tree how the material progresses from an elementary stage of raw material and purchased parts to the finished product.

Part Number	Part Name	Quantity
1001	Cart assembly	1
1006	Rigid casters	2
1007	Swivel casters	2

Fig. 6-4. A single-level bill of materials for Product 1000—Mobile Table.

Product 1000, Mobile Table

Part Number	Part Name	Quantity
1001	Cart assembly	1
1006	Rigid casters	2
1007	Swivel casters	2
1002	Table top	1
1003	Uprights	4
1004	Horizontal spreaders	2
1005	Diagonal braces	2

Fig. 6-5. A compilation of all parts from the bill of materials.

Figure 6-6 indicates that part numbers 1006 and 1007 are direct-purchased parts, but that 1002, 1003, 1004, and 1005 are shop-fabricated. The number in parentheses indicates the quantity of each part required. Examining the product structure tree (FIG. 6-6) reveals that the sheet steel tabletop, part no. 1002, is stamped from a sheet steel blank, the end-corners might be notched and the edges might be bent in one stroke of a die, or two operations might be performed progressively, depending on the equipment used.

The uprights and spreaders, parts 1003 and 1004, respectively, are cut from lengths of $1^{1}/_{2}$-×-$1^{1}/_{2}$-inch equal angle. The diagonal braces, part 1005, are cut from lengths of 1-×-1-inch equal angle. Further examination of the product structure tree reveals that parts 1001, 1006, and 1007 are higher-level components of product 1000 than are parts 1002, 1003, 1004, and 1005 because the former are higher in the product structure tree.

Inverting the product structure tree of FIG. 6-6 would give you the exact order, or sequence, with which the Mobile Table is to be fabricated; in effect, it would represent the entire work flow of the manufacturing plant, if the company were to fabricate only this one product. It would indicate, also, its complete material requirements in terms of exactly what parts are required, how many of each part, and in what order they are to be purchased, fabricated, and assembled.

Most companies, of course, manufacture more than one product; therefore, a series of product structure trees—i.e., one for each product to be manufactured—would

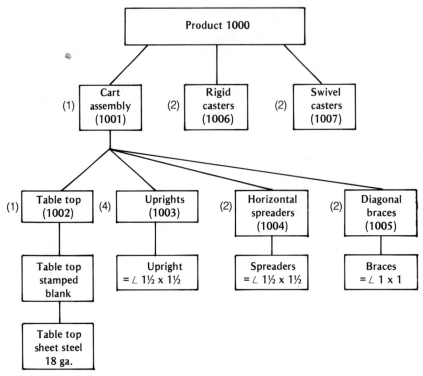

Fig. 6-6. Product structure tree for Product 1000—Mobile Table.

comprise and encompass the entire material requirements and flow of materials for the company. That is one of the salient reasons that MRP is of vital importance to the materials management organization—it serves as a road map for the company's entire manufacturing program.

Gross requirements

A part of the benefit of instituting an MRP procedure is the ability to determine the gross requirements for each product by exploding the bill of materials into a master schedule. The definition of *gross requirements* is the total component quantity comprising each end product. Figure 6-7 depicts the manufacturing and MRP network showing feedback elements as a closed-loop system.

In FIG. 6-8, the lead times required to purchase, or to manufacture, each part are clearly delineated by vertical dash lines. As you can see, the master schedule—having been exploded to the first level components of 1001, 1006, and 1007—indicates that the lead times for each part or component of the "exploded" product specifies the time intervals that are required. In other words, components 1002, 1003, 1004, and 1005 must be on hand one week (interval) before Product 1000 is required. When the time interval, in this instance "weeks," is combined with the master schedule by cutting up materials requirements in intervals of time, we have what is known as *time-phasing* of gross requirements.

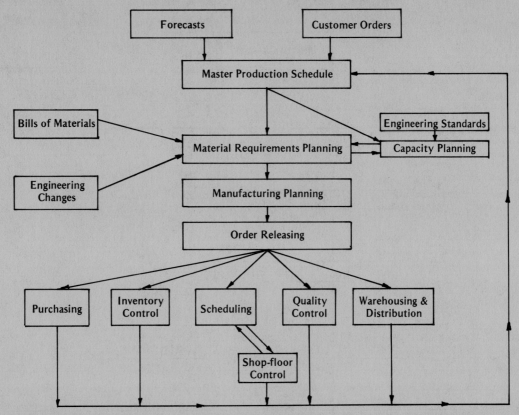

Fig. 6-7. Manufacturing and material requirements planning network showing feedback elements.

As indicated earlier in the chapter, the *planning horizon* is the time devoted to the entire scheduling period; therefore, the planning horizon for the master schedule must be sufficiently lengthy to extend from the beginning of the manufacturing cycle, where data is first entered into the planning procedure, to the end of the cycle, where the finished product results. Therefore, lead times for procurement and manufacturing, assembly, and so on must be considered so that all the materials that are required can be planned for and made available at the requisite intervals of time, right up to the point where the finished product is turned off the production line and packaged.

Figure 6-8 shows that the longest lead time element in the multilevel bill of materials for Product 1000 is as long as 12 intervals of time; in this case, 12 weeks. Other part number lead times are only 6 and 8 weeks in duration and are absorbed, time-wise, into the planning schedule.

The *cumulative lead time* indicates the earliest starting time that a component must be entered into the fabrication process or that a purchase order must be placed in order for the component to take its proper place in the manufacturing schedule. This starting date is the point at which the lowest-level component in the product structure tree must be ordered or fabrication begun. This sequencing of processes, or scheduling and pur-

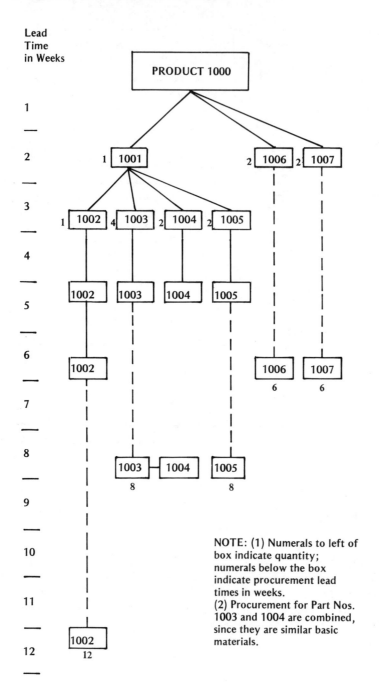

Fig. 6-8. Lead times given for Product 1000.

chasing functions, illustrates the need to maintain sufficiently tight controls on both the shop floor and the other planning activities in the materials management organization.

The complete integration and communication network between manufacturing and materials management is nowhere better revealed than in the MRP concept implementation. Feedback from the shop floor, machine loading planning, and all the other planning activities define what is to be the complete manufacturing schedule for any given time interval. It is the quality of both the shop-floor feedback and the planning activities that significantly determines how well manufacturing production is controlled—i.e., kept within time and cost constraints.

Unfortunately, scheduling methods vary considerably, as any manufacturing manager can attest, and while the basis of any manufacturing production program is its scheduling function, scheduling is more art than science. Also, process planning and scheduling will stand or fall on the basis of the quality of the people who fulfill these tasks. The training, maturity, and experience level of process planners and schedulers are important criteria for personnel in these areas.

The relative position of material requirements planning and the importance of feedback in the manufacturing enterprise are shown in FIG. 6-7.

Since feedback is an important element of control in the manufacturing plant, we can examine the master schedule of first-level components for Product 1000 (FIG. 6-9) and see the *lead time*, the time period in which each component is required. In this example, each of the components 1001, 1006, and 1007 must be available one week before Product 1000 is required in accordance with the time-phased determination for the product.

Part No.	Lead Time	Item	PAST DUE	Time in Weeks														
				1	2	3	4	5	6	7	8	9	10	11	12	13	14	15
Product 1000	1	Master schedule				150									200			
1001	12	Quantity required				150								200				
1006	6	Quantity required				300								400				
1007	6	Quantity required				300								400				

Fig. 6-9. Master schedule compilation of first-level components.

Gross requirements, which have been developed from the end product's master schedule and exploded bill of materials, sometimes might exceed actual requirements since, as it often happens, there might be a residual inventory of parts and components in stock or on order. Therefore, in order to determine net requirements, these quantities must be deducted from the gross requirements. As shown in FIG. 6-10, the number of pieces required for the final assembly of Product 1000 is adjusted for both "on hand" and "on order" quantities of the necessary parts and components.

It is worth noting that the lead times for "on order" parts are carefully adhered to, and where necessary the MRP concept will show where expediting or rescheduling activity is required. In addition, if certain parts are not required until later in the schedule, then minor adjustments can be made that might save carrying some inventories for weeks and sometimes even months. This is an additional cost benefit of MRP that makes it an attractive tool for materials management. Other advantages are that it enables planners to keep current with the real-time requirement for materials (parts and components), and it establishes priorities for meeting scheduled events in the master schedule.

Another use for MRP is to generate work orders and purchase orders, which are used by the planners to make certain that parts are available at the prescribed time to satisfy all of the net-requirement commitments that have been scheduled. Figure 6-10 illustrates this element of MRP. Planned orders have been generated in week 10 for parts 1006 and 1007 in order that sufficient quantities of the parts be available when required. The planned orders, in this instance, are similar to the lot size for the item.

It is important to note, also, that the quantities used for planned orders generated by the MRP system concept are not particularly a function of net material requirements, but are based on lot sizes which are established by a number of varying criteria, not the least of which is the cost of carrying the inventory. Thus, among the criteria are the following elements:

- *Cost of ordering* Ordering frequency should be minimized because it costs money to generate the paperwork and process the material from the supplier, through receiving, inspection, and storage for each shipment.
- *Packaging* Lot size sometimes can determine the manner in which parts and components are packaged.
- *Availability of resources* Since capital is tied up in inventory, the question often asked is how much money should be tied up in this type of inventory. If the number of different parts is large then large quantities of capital might be expended needlessly in large lot sizes.
- *Warehousing space* How much space is available for the required parts and components, and what sort of turnover is really necessary, or required, to permit the inventory to be warehoused?
- *Setup charges* Some parts require setup charges in the way of tooling and fixturing. Although it is usually not a major factor in the decision process concerning lot size, it might add an unwanted element of cost, especially when the setup requires new tooling and the like.

When you are considering lot size in an MRP installation, you must understand that the classic economic order quantity (EOQ) and fixed ordering quantity approaches

Time in Weeks

Part No.	Release of Orders	Status of Order Release	Past Due	1	2	3	4	5	6	7	8	9	10	11	12	13	14	15
1000	Lead time=1	master schedule					150								200			
1001	On hand 160 Lead time=12 Lot size=200	Required				150									200			
		On order	160															
		Net											190					
		Planned											200	200				
		Available		160	160	10	10	10	10	10	10	10	200	200	0	0	0	0
		Release	160															
1006	On hand 100 Lead time=6 Lot size=500	Required				300	500											
		On order	100															
		Net		100	100	⟨200⟩	300	300	300	300	300	300	300	⟨100⟩	⟨100⟩	⟨100⟩	⟨100⟩	⟨100⟩
		Planned							500				500					
		Available	100	100	100	⟨100⟩	300	300	300	300	300	300	300	400	400	400	400	400
		Release			⟨100⟩													
1007	On hand 400 Lead time=6 Lot size=500	Required				300												
		On Order	400															
		Net		400	400	100	100	100	100	100	100	100	100	⟨300⟩	⟨300⟩	⟨300⟩	⟨300⟩	⟨300⟩
		Planned							500				500					
		Available	400	400	400	100	100	100	100	100	100	100	100	200	200	200	200	200
		Release			300													

Fig. 6-10. Net requirements for the first-level components of Product 1000.

are based on the implication that demand will be continuous, rather than intermittent as it is in most MRP applications. A rundown of some of the lot-sizing methods will give you some idea of the large number of different systems that are available to the planning practitioner; however, not all of these techniques have the desirable characteristics that are required for the effective implementation of MRP. We shall start with the least desirable, then proceed on to the more useful applications:

Economic order quantity (EOQ)   EOQ is the lot size that minimizes the total cost of ordering, setting up, and maintaining an inventory. The EOQ makes the assumption that the use, or demand, for the part (or component) will be steady and continuous over a period of time. Because of this basic assumption, EOQ has virtually no advantages for use in an MRP program, since the attrition, or parts use, usually requires that parts be supplied in batches and intermittently, rather than in a continuous stream.

Least unit cost   The least unit cost is the lot size that produces the lowest setup cost, combined with the lowest inventory carrying cost per unit. The calculation of the least unit cost is based on the setup and inventory carrying cost for the number of parts required in one time period (month, quarter, or standard interval). Costs for the next time period's requirements are added to the first cost calculation, then cost is recalculated for the entirety. These calculations are repeated for requirements included in each successive interval of time:

$$x + 2, x + 3, \ldots x + n$$

Every successive total cost is divided by the number of required units in order to determine the quantity of units with the lowest unit cost. Unfortunately, this method of computing the lot size does not take into consideration the entire scheduling period, or the so-called *planning horizon*; therefore, it is not a desirable or acceptable methodology for use in the MRP system, although it has its application and should be considered as a viable alternative in some situations.

The Wagner-Whitin algorithm   The W-W methodology uses a dynamic programming model to calculate the best ordering quantity and interval, and the like, for the complete net requirements schedule. Since this methodology requires a vast number of calculations, it is usually not considered feasible for ordinary MRP use, despite the number-crunching capabilities of most computer-driven systems.

Fixed order quantity   The fixed order quantity method has limited applications, not only for MRP systems, but for other uses. The lot size in this instance is not determined empirically, but arbitrarily based on some established criteria. It might even be determined intuitively by a judgment factor. It is usually applied to parts whose characteristics preclude being taken fully into account by other lot-size methodologies; these characteristics might include tool life, storage problems, or near-future product design changes that would certainly affect parts and components.

Fixed-period requirements   This is probably the most simplistic of the lot-size determinations and, as such, can hardly be called a lot-sizing methodology. Nevertheless, it should not be overlooked, since it does have some measure of overall usefulness. In this so-called method, the practitioner would prespecify the number of time intervals—weeks, months, etc.—that each planned order would cover. In most applications, this method would be used when the parts requirements beyond a certain predetermined period of time are vague, or at least uncertain.

Least total cost (LTC) LTC is the lot size in which the cost for setup is equal, or very nearly equal, to the cost of carrying the inventory. In arriving at this premise, it is assumed that the summation of all setup and inventory carrying costs will be kept to the lowest possible amount, if these costs are equal. In order to achieve this summation of costs, it is necessary to analyze all the lot sizes required in the planning horizon—i.e., the entire scheduling period.

Part-period balancing (P-PB) P-PB uses basically the same formulation as the least total cost method for determining lot size, the only difference being that further total cost reduction is attempted by taking into consideration the wide swings in demand in the time intervals immediately before or after the time period in which an estimated lot size is planned. This method can be used, cautiously it is hoped, where seasonality or cyclical demand behavior is revealed to the practitioner with good results.

Lot for lot In this technique, parts are placed on order for each time interval in which demand is demonstrated, exactly as specified in the net requirements. The use of this method minimizes inventory carrying costs, but overlooks all the costs involved in ordering or setup. For this reason, lot for lot can be very effective as a technique for obtaining optimum lot sizes, but only if ordering and setup costs are low.

Period order quantity This method is quite similar to the fixed-period requirements methodology of lot sizing because the practitioner indicates a specified number of intervals each planned order should provide. The difference lies in the fact that the number of time intervals is estimated by ascertaining the EOQ and then annual demand requirements, based upon the market forecast telling how many end items will be required. The number of annual orders is divided by the number of planning intervals to determine the lot-size quantity per planning period. The final result combines the continuous stream methodology of EOQ formulation with the intermittent demand requirements of most MRP systems.

Although lot-sizing methods have considerable impact on the efficiency of the MRP concept, it is worthy of note that the release advisory, shown in FIG. 6-10, is one of the major outputs of the MRP system. The release instruction provides the notice to manufacturing or purchasing to fabricate or obtain a certain quantity in the time interval indicated. A compilation of these instructions indicates to the practitioner:

- What materials are required
- How many
- When the material is required

to fulfill the total end product (and spare parts and component) objectives of the master schedule. Thus, by pursuing and compiling the determination of net requirements through each level of the bill of materials for each product—that is, by exploding planned orders to generate gross parts requirements and adjusting them for "on-hand" and "on-order" quantities—the MRP plan is consummated. The net requirements quantities can be adjusted periodically as additional planned orders are required.

As you can see, the MRP net requirements schedule, or master schedule, can be fine-tuned continuously, and any additional customer orders can be cranked into the system schedule as the need rises; this is true of special orders, firm orders, and the like. Therefore, the mechanism to override the system is incorporated into the MRP methodology to give it the flexibility of a dynamic programming device.

Since FIG. 6-10 illustrates net requirements for the first-level components of Product 1000 manufactured by XYZ Mfg. Co., a little stretch of the imagination would indicate that all the parts and components of Product 1000, from first-level components down to lower levels for all the products manufactured by XYZ, can be netted out in this fashion. It is interesting to note, also, that lot sizes might depend not only on Product 1000, as our illustrative example indicates (FIG. 6-10), but on a combination of lot sizes where parts of one end product are also used in other products.

For parts and components that appear in several levels of bills of materials, the gross-to-net computation must be carried out to the lowest level at which the part occurs to make certain that all gross requirements have been included and compiled prior to explosion. A low-level code assigned to each part would indicate the lowest level in which it would appear in any bill of materials. An example of the coding is shown in TABLE 6-1.

Table 6-1. Part Levels in Production.

Part No. Level	Quantity	Part Name
.1	1	Cart assembly
..2	1	Shelf
...3	1	Sheet metal blank
.1	2	Swivel caster
.1	2	Rigid caster

The assignment and identification of a low-level code for a part indicates that the compilation and explosion of parts and components should be held in abeyance until all gross requirements from higher levels have been identified and tabulated. If, for some reason, the explosion of gross requirements is carried out in haste, rather than performed in its proper sequence, the result will be the understatement of net requirements at each level of the exploded bill of materials. Unfortunately, this outcome results in inaccurate lot sizes and incorrect or inadequate safety stock quantities.

It is for this reason that meticulous attention to detail is required, and personnel who are working in this area should be chosen for their ability to work with figures. Not everyone possesses the innate ability to deal with numbers; therefore, it is suggested that aptitude tests be administered to potential candidates for these positions. Some people, even very intelligent persons, cannot count with any real degree of precision, as I have discovered in supervising wall-to-wall inventory procedures.

The necessary review and the cycle of computation and adjustment must continue through all levels of bills of materials for each end product. When multiple-use parts and components appear in a bill of materials, every previously computed gross-to-net calculation must be repeated, every lot size adjusted, and all safety stocks refigured. In addition, since the net requirements for the particular part have changed, the gross requirement for all lower-level components dependent on that part's net requirements must be recomputed.

Application of MRP

In the previous section, the net requirements of the master schedule were illustrated as a computer-prepared document (FIG. 6-10), which was developed for the first-level components of Product 1000. It was suggested that by carrying this concept further, in a similar direction, all of the multiple levels of the many bills of materials representing the complete product line of the XYZ Mfg. Co. could be exploded and represented on the master schedule. The purpose was to give as complete a set of operating instructions as possible so that the plant in its several departments could take timely action to see that all the material movements required could be made in a time-phased environment to achieve the targeted objectives of the company.

This then is the basic output of the MRP concept: to schedule lot-sized orders in an economical manner to satisfy net requirements for parts, components, and raw materials over a fixed planning horizon. For this purpose, the master schedule shown in FIG. 6-10, when expanded to cover all of the bills of materials, is both informative and instructive because it becomes an operating tableau of the enterprise when the following data is supplied:

- What is required and when
- What is on order
- What are the net or residual parts, components, etc.
- What is planned and when
- What is available and when
- When parts and components are to be released

The MRP output then provides a set of operating directives for the production control, inventory control, and purchasing departments, which can take any or all of the following forms:

- It can release planned orders to purchasing or production control
- It can cause purchasing and production control to issue expediting orders
- It can reschedule existing orders to different periods
- It can change the quantity, or cancel existing orders

Updating MRP systems   It is possible to categorize MRP into two types of systems, depending on the methods used for updating net requirements: regenerative MRP, and net change MRP.

In the *regenerative MRP update*, there is a periodic explosion of the entire master schedule using multilevel bills of materials for all end products to create new planned orders and instructions for actions to be taken in order to maintain the master schedule on target. In the *net-change MRP* philosophy, only specific changes to the master schedule or to the status of individual parts, materials, or components are exploded in particular bills of materials.

Regenerative and net change systems give virtually the same results; however, there are important differences in methods. The frequency of planning, making changes, and planning again is relatively high for the net change system. On the other hand, planning frequency is fairly limited when the regenerative method is employed. When net change is used, planning response to changes is faster, although net change systems generally require followup to periodically purge the system of bad data. If the company's data processing capabilities are relatively weak, then this is a fairly good reason not to embark on an MRP net change system for updating the master schedule.

Are safety stocks necessary? The principal reason for having safety stocks in an MRP environment is to provide a buffer for unanticipated fluctuations in demand. The necessity for having safety stocks indicates, primarily, that the MRP methodology is not being used properly, since it should be one of the inputs provided for in the planning required for establishing the master schedule, especially when independent demand items are being considered.

When the safety stocks concept is placed into perspective, it would seem that tailoring the plant to Just-In-Time manufacturing would resolve this problem. Nevertheless, because of the futility involved in obtaining Japanese J.I.T. efficiency in the light of the differences in work ethic, temperament, and union-management relationships, we can isolate the requirement for safety stocks based only upon parts and components with dependent demand and where supply is somewhat uncertain, as in the case of only a very small percentage of a company's purchased parts. Therefore, provision for safety stock can be placed into the MRP system prior to the instructions necessary for the release of orders.

In this strategy the normally derived order quantity is placed on order, but may be received earlier than required. With this technique the on-hand balance will include the safety stock, which is equivalent to the number of time intervals in advance of the period in which it is to be consumed, and the aggregate-on-hand quantity will be equal to the average use per time interval multiplied by the number of intervals. In companies that do not have an active J.I.T. manufacturing program or who pay lip service to this concept without actually having a cadre of suppliers who can deliver parts within the hour or the day of use, the MRP practitioner might feel the need to maintain inventory levels that are sufficiently high to avoid stock depletion caused by production delays, late deliveries, and parts not up to specifications, which have been placed on "hold" by the quality control inspector.

Safety stocks for parts, components, and raw materials can be factored into the MRP system by reducing the quantity of safety stock from on-hand quantities or by increasing the quantity of safety stocks in the gross requirements. Using either of these two methods causes an upsurge or increase in net requirements. Although this might appear to be a solution to the problem, it can return to haunt the practitioner because, when used with MRP methodology, it might cause the system to disgorge overstated requirements, thereby distorting order priorities and throwing the master schedule out of synchronization.

MRP implementation   MRP is, by and large, a philosophy of operating a manufacturing enterprise, rather than a mechanical system or device that is plugged into a computer network like a canned program. The same thing applies to J.I.T. manufacturing. It is not a concept that is easily assimilated; the whole mind-set of a company must be changed to accept the philosophy, and a great deal of preliminary work must be accomplished by a dedicated staff. In this manner MRP parallels the extensive preparation required for installing a J.I.T. system. Also, it is a companywide approach that is relentless in its adherence to certain principles and ground rules, since it usually requires a radical change in the company's approach to manufacturing.

Preparatory activities, prior to installation of the MRP system, require a complete analysis of the company's strengths and weaknesses in such important areas as identifying customer demand and forecasting based on marketing information and strategy; gauging the marketing department's ability to detect shifts in customer demand and its reliability to determine the time interval and the quantity of such changes in customer intentions. In addition, customer service goals are important considerations. For example, how soon after receiving a customer order or after determining that a product will be fabricated should a customer receive the items or should it be ready for distribution?

Response times for customer service, spare parts reaction time or readiness, and the like pose other restraints and conditions with which the MRP team must be completely in accord. Also, apropos of customer service, the MRP team should consider the number of stocking points for inventory required to provide adequately for the lowest possible response time for both end product and service parts. Then, again, there is the question of in-plant inventory location and quantity and the necessary investment capital required to maintain all inventories regardless of whether they are in-plant or at customer service facilities. Above all, the team must define the level of customer support to be achieved, and ensure inventory record accuracy.

In addition to all of these items, actual in-plant operations must be analyzed carefully. For example, in manufacturing it would be best to start with the paperwork and communication elements, and then proceed to the physical aspects of operation. This is where the materials management department must take the initiative and verify the accuracy of the current bills of materials, updating them where necessary so they agree with all of the model changes and design modifications (and engineering changes) that have accumulated during the life of the product. This is no small task, but it is essential to the success of the MRP team's assignment.

Next, the materials management department must analyze work-process flow in conjunction with the planners and the manufacturing engineering staff. During the review of engineering standards and work performance standards, the industrial engineering department should work closely with manufacturing and materials management to establish a frame of reference, which should permit subsequent evaluation of the efficacy of the MRP program as it progresses.

This is a relatively important phase of the preplanning function for establishing MRP in the plant because you want to be able to look back periodically and see how well you are doing. Without the historical benefit of these benchmarks, you tend to lose the perspective of past performance. Then again, if real improvements in manufacturing and materials-handling operations are being made, it helps the program to gain companywide acceptance and sustain the support from top management, which is required for the continuance of the program.

Another phase of the preparatory work is the analysis of purchasing performance, which is directly in the heart of the materials management organization's work. Such aspects as purchasing volumes per buyer, lead times required, qualification of suppliers according to response time and status of due-ins, and in general the overall performance of the purchasing department in support of manufacturing schedules should be reviewed and documented carefully. If the performance of purchasing has not been as effective as it could be, then organizational changes might be required.

There is a good deal of difference between collecting data and analyzing it. In reviewing data that has been so diligently collected it would appear necessary to have this review conducted by persons who are familiar with materials management systems in general, and with the company's extensive manufacturing operations. Because it is extremely difficult to remain completely objective when analyzing one's own operations, the task force that is given the charter to install MRP should be composed of a fairly high-level group, with at least one representative from each of the concerned departments (which have just been mentioned). Also, since costs are the primary concern of the financial and accounting department, member(s) from this organization should be included in the task force, as well. Sometimes it is necessary to draw upon management consulting expertise from outside the company's ranks to keep the necessary objectivity in clear focus.

The principal areas of concern for the task force revolve around three major areas, as follows:

The accuracy of inventory records   Another way of stating this area of concern is to call it the *precision* of inventory records. Precision, and a high degree of it, is an absolute requirement for the successful implementation of an MRP program. MRP, by and of itself, cannot cure inaccuracies in inventory; quite the contrary because it will exacerbate the negative effects of these errors. As an example, if inventory quantities are too high, there will be a failure to order materials when required. Conversely if the records are on the negative side, an oversupply of items will be ordered. Subsequently, failures of this type will produce erroneous gross-to-net requirements in the master schedule, and the MRP system will collapse. If inventory record-keeping is not very precise, then expediters and other inventory-control personnel will begin to handle shortages outside the system. This procedure would have a snowballing effect, which would ultimately destroy the whole purpose of the efficacy of MRP as a manufacturing tool.

The task force examining inventory records should review a broad spectrum of the parts using a valid statistical sampling methodology in order to determine the reliability percentage of the records. A reliability check on relatively low dollar-value items should be within ±5 percent of the numbers indicated on the inventory records. Scale-counted (counted by weight) items should be as close as ±2 percent to be considered acceptable. The main thrust of the inventory accuracy task group should be to improve methods and record-keeping so that these minimum standards can be attained, and maintained. Cyclic inventories are now largely supplanting wall-to-wall inventories, and bar-coding techniques and equipment can be employed to minimize counting and record-keeping errors.

A number of other techniques also can be used to promote the reliability of inventory records. In the first place, unauthorized persons should not be permitted access to the warehouse portions of the plant. In some manufacturing operations, some self-proclaimed expediters will leave the manufacturing areas and pick up parts and components they need to complete an assembly without considering the havoc they are causing in inventory control, simply because they are frustrated by delays in parts supply or the material-control personnel have delivered an insufficient number of pieces to production. This "self-help" problem must come under the close scrutiny of materials management because it is an indication of a faulty materials supply bottleneck. A temporary solution, although having long-term benefits, is to tighten up on the paperwork procedure of the warehouse and prohibit the access of unauthorized personnel to the storeroom.

Other ways in which errors creep into inventory records is through the absence of adequate receiving department procedures. It is necessary that procedures be formalized so that nothing is released from this department without careful documentation. Quality control inspection rules should be such that "holds" on materials and "set asides" should have such stringent controls that parts taken by inspection are either replaced or otherwise accounted for. Also, locked storerooms help prevent pilferage, and this philosophy applies to QC container opening. Every container that is opened for inspection should be resealed and stamped "Opened for Inspection."

It is by tightening up controls such as these that, ultimately, will achieve inventory accuracy. This leads to another area: the use of adequately designed paperwork forms to be used with the receipt and issuance of materials. Certainly, this area should be reviewed carefully by the MRP task force, for if the mechanism to track materials is lacking, then the system reliability cannot be sustained.

Bills of materials The composition of bills of materials has the undeniably important task of describing the contents and structure of the end product. It is for this reason that the MRP program requires complete accuracy in this regard, which means that all bills must be brought up to date with deletions and additions—i.e., engineering and design changes, and changes in quantity. Helpful to the MRP plan, also, would be the identification of the level of use of each part and a note as to whether or not it is used in more than one level of manufacturing. The use of a low-level code for each part would inform the MRP practitioner of the lowest level that the part is found in any of the bills of materials. This has future MRP significance because the use of a low-level code would decrease processing time and eliminate many of the errors that might slip in during the explosion for regeneration or the net change procedure by eliminating the repetitious explosion and order planning of multiple-use parts, components, or subassemblies.

In conjunction with the task force's review of the bills of materials it would appear practicable, at this time, for the group to review the way in which engineering changes are handled, especially if there is conspicuous evidence from the examination of bill-of-material accuracy that design and engineering changes are not effectively being introduced or communicated to the drafting department, or the group responsible for maintaining the currency of these lists.

Master schedules   Since MRP represents a very powerful adjunct to the materials management philosophy, it is important to emphasize the part that the master schedule plays in the MRP concept. The master schedule is the very heart of MRP, and as such it must be sustained with accurate data that will make it as complete as possible so that it adequately represents the manufacturing objectives of the company. In the course of *purifying*—that is, extracting erroneous information from the master schedule—it would be well to observe the GIGO adage from the computer world: i.e., garbage in, garbage out.

One of the purges to be performed on the master schedule is to weed out the production of nonessential or superfluous parts that might exceed the actual capability of the plant. Another area of concern to the MRP practitioner is the production of end products that do not appear on the master schedule. Such work has the upsetting effect of distorting machine loadings and will in time, play havoc with any attempt to adhere properly to a time-phased net requirements schedule.

In addition to these defects, another potentially dangerous condition is the reliance upon planning horizons that are of insufficient duration to permit the proper lead times for both fabricated and purchased parts and components that will accommodate items with exceptionally long lead times. Other difficulties that might arise with the master schedule are the ways in which schedule changes are handled, especially when there are lead times in the involved changes that cannot be absorbed readily by the schedule because of shortened performance intervals.

Being forewarned is being prepared to take all of the corrective measures required to place meaningful data into the MRP master schedule during the preparatory preplanning stages of the MRP installation. Therefore, it is necessary to identify and correct all the weaknesses and potential problems in the plant's manufacturing and materials management operations as a preparatory first step. It cannot be expected, of course, that the implementation of an MRP system is a panacea that will solve all of the plant's manufacturing problems; however, the symptoms that will be apparent if there is a real need for the MRP installation can be condensed as follows:

1. If the manufacturing department experiences frequent stock-out situations, which in turn produces a periodic frenzy of expediting actions both in manufacturing operations where a great deal of overtime labor hours are expended, and in purchasing where overtime may, or may not, be necessary.
2. If inventory dollars start increasing at a too rapid pace, indicating that inordinate quantities of safety stocks are being used as a cushion for poor planning and inventory, or production control practices.
3. If customer complaints concerning poor deliveries are on the increase.
4. If either direct or indirect labor costs are growing at ratios faster than the output of the plant can support.

The practitioner who is attempting to justify the installation of an MRP system should be well versed in areas of manufacturing engineering and inventory and production control in order to present the benefits of MRP convincingly to top management. In the first place, if there is a certain dissatisfaction that upper management might have with the plant's performance—for instance, a return on equity that is somewhat disappointing, or that has experienced slippage over the past few years—then they might want to revamp their manufacturing operations after an exhaustive review of all manufacturing systems. The manufacturing analysis might point out the benefits that can be obtained by the installation of MRP.

The installation of an MRP task group would have as its charter the following targets:

- Reduce inventory costs (freeing capital and permitting it to be used to reduce debt, or to earn interest)
- Improve customer service
- Reduce manufacturing costs
- Reduce indirect labor costs
- Improve purchasing performance

Installing an MRP system

The logical place for the initiation of the MRP philosophy is in the materials management department, or in its predecessor organization. Some interested and alert individual or group of individuals must respond convincingly to the malaise expressed by top management on the question of how to improve the company's return on equity.

If it should be decided then, after top management has been thoroughly convinced of the potential benefits of MRP, that the materials management department is to provide the guidance and supply a good deal of the manpower to get the show on the road, what should be the first steps? From the previous section, we know that there are certain preplanning steps that must be taken by the MRP task force prior to the actual, full-scale implementation of the system. In the first phase of implementation, an MRP task group should be selected and manned by representatives from the following departments:

- Manufacturing engineering
- Planning and processing
- Accounting
- Data processing
- Inventory and production control
- Materials handling
- Marketing
- Industrial engineering
- Purchasing
- Materials management

A project leader should be selected, preferably from the materials management department, and the endorsement of the project and leadership should be made at least at the vice-presidential level, if not from the CEO of the company. The task group should be a dedicated entity with full-time responsibility for the installation of the MRP methodology. Otherwise the success of the program might be questionable; at the very least, it will take longer to install, and might be unintentionally undermined by a lack of attention to detail for the complex relationships that exist between many of the departments composing the company.

As a second step, the task group should decide whether or not a packaged computer software program for MRP should be purchased and modified to suit the company's individual requirements, or whether the company should attempt to develop its own in-house computer software programs to drive the MRP system. The purchase of MRP software will, in all likelihood, hasten the installation task. In today's marketplace there are a number of computer manufacturers and software suppliers that can furnish a large variety of analysis routines that enable the user to simulate and examine the effects of changing the ordering system.

In comparing the alternatives of commercially available software to using in-house computer personnel for development purposes, several of the following underlying areas of concern should provide answers. For example, it should be determined if the company's existing information-handling systems are adequate to maintain the kinds of data required in the new system. It should also be ascertained if the systems for information/data gathering are currently being used in the manner in which they were originally designed or if they have been modified or otherwise misused in the attempt to fill new needs, as sometimes happens when changing factors within, or external to, the company dictate these needs. The members of the task group from the data processing department can, of course, provide valuable insights into the decision to purchase available software once they are familiar with the objectives of the MRP system. And, in evaluating the cost advantages or disadvantages of the two approaches, they will have important input into this process.

Since the task group will report progress, periodically, to top management, there should be an early review of the cost benefits of the new system, in as realistic a framework as possible. Next there should be a clearly defined statement that will contain the objectives and anticipated results of the new MRP installation. There should be, of course, periodic updates of the progress of the task force, but in the initial review of objectives there should be some mention of these major categories:

- Verification of all bills of materials in order to eliminate errors
- Upgrading of inventory records to at least a 95 percent, or better, degree of precision
- Provision of inventory records with realistic and accurate lead times and order quantities
- Development of a master schedule that is complete and realistic for all parts, components, and end products

Concurrently with this phase of the MRP installation should be the companywide training program that will prepare all of the employees for the new methodology. The new look at operating the company will involve everyone from the top echelons down to the lowliest stock clerk. The main emphasis will be production and inventory control innovations and the MRP way of looking at scheduled events. There should be formal training programs for all employees so that the methodology and philosophy of MRP is completely instilled in them, and they should be encouraged to develop a sense of responsibility for the success of the program which will increase the success and profitability of the company and contribute to further job security. In this regard, the system's success or failure rests with the employees, and the more they learn (and know) about the MRP technique, the quicker and more cost-effective will be the installation.

The third phase of the MRP installation, after computer software is in place, would be to start with a single end product, or product line, with a few multiple-use parts and components as possible, and begin processing it as a pilot operation. It is important that the processing (routing) sequence for each part be reviewed to make certain that it is the current and most effective method before converting it to the computer program. As a result, the part drawings also must be reviewed to update and introduce the latest design and engineering changes, so each drawing in the product line must also be reviewed completely for configuration, fit, and finish. Part dimensioning and group technology aspects should be reviewed at this time to take advantage of any possible benefits.

Sometimes this aspect of the MRP approach might open up a can of worms, so to speak, if the company has not progressed to the point where manufacturing engineering time has not been invested in exploring and benefiting by *group technology* (GT). In GT a family of parts for all end products are grouped for economies of processing on like machine tools. It would be best to defer this concept until such time as the MRP program is well established, since it would be too disruptive to include this program conjointly with the MRP installation, to say nothing of the additional engineering and computer time involved.

As far as the MRP logic is concerned, formal capacity planning and input/output control should be installed at this point, and shop floor controls should be established to make certain that there is feedback to determine if what is happening in the manufacturing processing is following the master schedule as planned.

When the prototype product line is running satisfactorily on the MRP system program, then related groups of products should be added to the program. A word of caution is advisable at this juncture: engineering changes, which can come with astounding rapidity in some plants, should be held to a minimum. I am familiar with one well-established machine tool builder who, customarily, had over 300 engineering changes weekly. A rash of changes of this magnitude would be very difficult for a fledgling program to handle.

The MRP system, while managing to cope with engineering changes, might at the same time result in the diminution of orders to suppliers. Since suppliers also must be educated to the MRP system and the reduction of the frequency and quantities of orders that might transpire as a result of the efficacy of the MRP logic, it would be well for the material planners and the purchasing department to reduce orders in small increments so as not to upset the vendor population.

During the initial phases of installation there should be an ongoing effort to ensure that there is complete bill-of-material accuracy, and inventory record precision should be continuously verified by increasing the number of cyclic inventory counts.

It is evident from the foregoing discussion, the manner in which the MRP procedure sharpens and fine-tunes the manufacturing, processing, inventory control, planning, purchasing, and related departmental operations, and places enormous responsibility on each to fulfill the requirements of the system. It is this continuous reliance on all members of the structured organization for their top effort that contributes to the profitability of an MRP installation.

The same basic techniques that have been described in this section can be applied to all the MRP systems described earlier in this chapter. It is, indeed, fortunate that the design parameters of each of the systems can be customized to suit the individual characteristics of the manufacturing environment in which they are employed. Input/output, lot sizing, exception rules, and the like can be customized to fit the circumstances and changing prospects of the company and the marketplace. For example, whenever there are strike hedge strategies to be pursued, material shortages or gluts, rising capital costs to carry inventories, and such, then it is possible to vary the system's operating parameters to conform to the moment's exigencies and still have a workable, usable system.

Although it is easy enough to change operating parameters, the ease with which it can be accomplished should be a sign that too much change might abort the overall effectiveness of the technique. Therefore, before any change is made, its long-term effect on the system should be considered carefully.

Whenever changes in the operating parameters are to be made, it would be expedient to pass the changes along to the original task force members for comment before any action is taken. After a consensus is obtained, the changes, or any modification thereof, should be distributed throughout the company so that everyone who needs to know will be kept current on the latest modifications to the system. It would be prudent also to have MRP refresher updates periodically to make sure that everyone is up to speed on the latest MRP developments. New employees also should be invited to these updates as a means of indoctrination into the methodology.

Another method for strengthening user enthusiasm for MRP is to have top-echelon managers of the company invited to these updates. At such times, a few cogent remarks from a company vice-president will lend support to the MRP program and serve as encouragement to all employees.

Material resource planning (MRPII)

In the decade or so that MRP has enjoyed its existence, there have been a number of modifications to the technique, which have encompassed more and more of a company's operating divisions and have focused increasing attention on the part that finance plays in the overall operation of the manufacturing-distribution complex. One of these transitional phases was known as *closed-loop MRP*.

The way in which closed-loop MRP developed is that MRPI practitioners became aware that the systematization of plant effort resulted in a very effective scheduling method. It made no difference whether the plant's end product was made to order or made to stock, or whether it was comprised of assembly components or end-product assemblies, the material requirements were capable of being derived from the master schedule. Therefore, in order for the master schedule to have validity, the capacity requirements had to be valid, also!

Thus, closed-loop MRP produces feedbacks that provide answers to problematic areas, such as:

- The quantity of end product the plant is capable of producing
- The on-hand inventory and available manufacturing capacity that exists
- The parts, components, and materials that must be obtained in order to produce the given volume of end product

By providing this information, the system is then capable of estimating whether or not the production volume forecast is feasible. In this manner, a work schedule can be produced for each factory work center, as well as a weekly master schedule and the accompanying directives that will drive the system. It is then possible, through feedback control, to monitor the capacity plan and determine how closely actual production compares with planned requirements.

The next step in the transitional process was to proceed from closed-loop MRP to *manufacturing resource planning* (MRPII). The result of systematizing the closed-loop approach provided a means for planning and tracking all the resources of the manufacturing plant including, but not limited to, the following:

- Financial
- Marketing
- Engineering
- Sales
- Manufacturing
- Purchasing
- Inventories

MRPII uses the methodology provided by closed-loop MRP to evolve financial data, which is generally expressed in dollars. The amount of dollars involved is generated from the material requirements plan, in which a specific number of end-product units are comparable to dollar quantities at various cost levels. These dollar quantities are called *standard dollars*, in economic verbiage. However, the total dollars representing total end-product assemblies must be converted, or exploded, into their component parts in order to obtain a projected inventory figure. This number represents the proper amount of inventory to have "on-hand," or "on-order," with which to produce the quantity of end-product parts, or components, which have been indicated as required on the master schedule.

If the master schedule has been assembled with a high degree of precision, then performance can be gauged on a daily, weekly, or monthly basis from the feedback of the system. This feedback provides the MRP practitioner with the capability to compare the reality of the shop floor with the master schedule so that not only can he or she determine the effectiveness of the installation, but can make adjustments, where necessary, to bring the two into synch.

When an MRP system is working as well as it can, it is relatively easy to obtain standard hours for each manufactured part and to convert these hours to capacity requirements for each of the shop-floor work centers. Extrapolating these figures permits the practitioner to obtain labor hours and costs, and to resolve make-or-buy decisions. Inasmuch as all operating units within the company are involved in the MRP process, an added advantage of MRP is that all of the departments can be working with the same set of numbers.

Installing an MRP system usually requires from 18 months to 2 years of very intensive work by a dedicated group, since it is a complex installation involving a considerable amount of computer programming. As indicated earlier in this chapter, there is also a considerable change in the way in which almost everyone in the company must operate.

It should be clearly understood at the very earliest point that MRPII does not focus primarily, or solely, on changes to the company's computer database. It is not merely a data processing type of project, but it is mainly a companywide behavior-modification program in which almost the entire company is involved in some way or another.

At one plant in my experience, almost 40 percent of the suppliers were past-due in their shipments. Since this was considered normal in the fairly slipshod operation, it had become an accepted practice for the purchasing personnel to disregard the delivery requirements of the purchase agreements. No one in the department seemed particularly disturbed by this fact, and their apparent lack of concern stemmed from the erroneous inventory figures, which were distorted by either inflated, or skewed, numbers. MRPII put an end to this complacency, since it demands a high level of inventory precision as a result of the requirements (from bills of materials) of the master schedule.

Although it is desirable that any company adapt the full-scale MRPII methodology for the full cost benefits accruing to the system, it is possible to implement any reasonable portion of the methodology that will heal ailing portions of the company.

Distribution resource planning (DRP)

Distribution resource planning should be of considerable interest to the materials management department, primarily because its major use is as a scheduling methodology. When DRP is employed by materials management, it is possible to replace the "reorder point" method of providing inventory for a distribution/warehousing group (which issues customer products). DRP is a method that permits the receipt of a large number of products from a multitude of suppliers and manufacturing sources wherein each product, part, or component can be tracked, shipped, received, warehoused, and order-picked while achieving a high level of customer service of 95 percent or better.

The DRP system can interface with any of a company's inventory tracking and scheduling systems in such a way that feedback required in the comprehensive MRPII configuration can be provided to complete the systems data network. Information obtained from the DRP source is useful in preparing the manufacturing master schedule on the production side of plant operations, but its chief contribution is the data it provides for transportation planning and for the development of such shipping information as weights, size (cube), number of pallets required, and the like. All of this data can be entered into the central data bank (with proper coding to make sure that the information can be retrieved), where it serves as a powerful adjunct of MRPII and the computer-integrated manufacturing (CIM) methodology, if and when this concept is used by the company.

Capacity requirements planning (CRP)

The orderly flow of materials through the plant is the primary objective of materials management. When the movement of materials flows from one stage, operation, or process to another, ultimately into the distribution channels of commerce, the materials management group can consider they have done their task well. Unfortunately, in the manufacturing environment, as so often happens, many operations are conducted with so many loose ends that activity in the plant tends to run on what is known as the *short sheet*—the list of parts that are urgently required in order to complete the assembly of subcomponents, or even the end product.

More in-house turmoil and consternation is created by the short sheet than by any other single shortcoming in manufacturing practice. It usually falls upon the shoulders of the shop foreman to obtain all the parts of the final product assembly when the production control routine becomes unraveled. The breakdown that is usually experienced might be the lack of some small, relatively insignificant part that is not even a large-volume production part failure. Nevertheless, the lack of any one part can become a major dislocation to the orderly flow of semifinished assemblies. Components and sub-assemblies, or assemblies in the final stages of completion, must be sidelined and storage can become another vexing problem to the harassed shop foreman who finds himself in this situation.

Short-sheet operations usually create a host of other problems in the factory, a major result of interrupting orderly work flow in the assembly areas, which creates morale problems not only in this department but in the fab shops as well, where jobs on machines must be removed to be replaced by the expedite orders. The worker who is pulled off one job setup and told to run another rush job is inclined to feel that management is inept (as it certainly is, to permit this flagrant violation of work management), and the loss of respect that department foremen strive so hard to acquire is lost in the last-minute hysteria to keep the end product moving out the door. Both the employee on the line and the foreman become relatively disgruntled employees, and their resentment can be manifested in many strange ways in the manufacturing environment.

In addition to the morale problems that exist with the short-sheet procedure, there is the inevitable loss of quality, the decrease in productivity levels (because, remember, we have introduced another setup, transportation, and materials-handling operation into the overall productive time period), and the delay in delivery schedules. By and large, the reputation of the company might suffer from too many of these so-called "emergencies."

As a psychological aside to this discussion, in my experience there have been a few shop-floor foremen and middle managers who thrive on this type of crisis management and feel that they are being superb managers by the way in which they dash into the breach and obtain the shorted part with a combination of yells, screams, and threats, which endear them to all their colleagues. One gentleman even had the nickname of

"Short-Sheet Charlie" because he was so good at storming into the neighboring machining area, making loud noises, to obtain his lacking parts.

One of the exasperating facts about this philosophy of operation is that it is virtually unnecessary if the proper planning and follow-through mechanisms are provided. Thus, in examining what constitutes good shop-floor control it is discovered that in order to make a good-quality product on time there must be the proper coordination of the correct number of parts, proper tooling, good scheduling, and machine loading—and, capacity requirements planning. Each and every functional area of the company must be responsive and demonstrate the capability of working toward this goal. The objectives are quite clear: to have enough capacity at each work center and to be working at the correct jobs at each work center.

Good information is required to adjust to the continual barrage of changes in requirements that represents the dynamics of factory operations, and that is the reason that the ubiquitous computer terminal has made its appearance on the shop floor. When all facets of manufacturing are coordinated, shortages due to poor planning and scheduling will virtually disappear, and the extra cushions of inventory will no longer be necessary to prop up a poorly conceived system. Productivity will improve, the cost of manufacturing and materials handling will decrease, the inventory turnover rate will improve, and morale, in general, will improve—as well as customer satisfaction.

In the manufacturing environment, capacity requirements planning (CRP) can be accomplished only when production planning, materials requirements planning, and the master schedule have been developed. The production plan is the outcome of marketing studies and top-echelon decisions regarding what and how many of the company's end products should be produced. This plan is a crucial aspect of plant management and, as such, should be done by professionals since it takes the plant from the present to some distance into the future. Since it is not an analysis that is strictly quantitative, it must constantly respond to changing market conditions. The quantities, types of products, and models will change from time to time as marketplace demands fluctuate; however, once the production plan receives its final stamp of approval, all manufacturing departments—in coordination with purchasing, engineering, finance, and materials management—are committed.

The master schedule is a reflection (almost a mirror image) of the production plan. It breaks down the production plan into the number of parts of each product that are to be produced or purchased, and it tells when the parts are required. As part of the MRP system, CRP indicates machine loading and routing functions. The data for these functions describe the operations that are required, the sequence in which they are to be performed, and an estimate of the standard hours to set up and run the parts. A summary of engineering hours is calculated, and allowances are included for any operations that are not standard. Hours are then computed, including transportation times between operations and work centers, waiting times, and related times. Work centers are identified so that capacity data can be summarized realistically.

In this frame of reference, a *work center* can be considered as a group of similar or related machine tools, or as a group of workers with identical skills, as for example, grinding operations. The projection of capacity for each work center can be extended as far into the future as the master schedule's terminal point.

Notes

1. Micro-MRP, Inc., Century Plaza One, 1065 East Hillside Blvd., Foster City, CA 94404. 415 – 345 – 6000. Fax 415 – 345 – 3079.

2. Since the planners know how many labor hours they have to work with, it sometimes happens that production for stock will assist in rounding off labor-hours. If differences in ratios of production to available labor-hours become too great—i.e., when business falls off—it presages a dislocation in labor effort.

3. The definition of *finished product*, in this context, is the finished, assembled unit that is offered for sale to the public, is self-contained, is usually composed of several parts and/or components, and is differentiated from a spare or service part by the fact that it is the original equipment, or product, sold.

7

Flexible manufacturing systems and the materials management link

Introduction

To understand the relationship between materials management and the type of manufacturing automation known as FMS, or flexible manufacturing systems, it is necessary to realize that manufacturing automation, in general, is composed of such basic elements as (a) machine tools to do the work, (b) materials handling to get the work (parts) to and from the machines, and (c) the controls to make this happen. Edison and the early automobile manufacturers, such as Ford and Sloan of General Motors, ushered in a wave of mass-production techniques in manufacturing as World War I was being fought, and the need for productivity enhancement led to the development of machine tools that were customized for specific tasks.

In the beginning of this machining revolution, special machines were used to perform simple operations and were driven by line and shaft power. A straight line shaft was run overhead and each machine tool was connected to its own pulley on the overhead line with a leather belt. At this time, every factory using machine tools looked like a forest of leather belts! Each special machine used for turning, drilling, milling, etc., had a single spindle and performed a single operation on a workpiece, which was fed into the machine by an operator and then passed on to the next machine in line by means of a nonpowered gravity roller conveyor.

Since output of line-shaft operations was limited by the time it took to load, machine, and unload each piece and then transport it to the next operation, it was a natural development to the individually electric-motor-driven, multiple-spindled machine tool. With this innovation, the straight-line production layout was superceded by the disposition of machine tools wherever they were required to perform a number of operations, and manufacturing plant layout was vastly improved. The multiple-spindled machines were capable of driving several spindles by means of a gear train, which allowed several operations to be performed by one machine. Although the machining times did not improve significantly, the fact that more machining operations could be performed simultaneously within one time cycle, and one operator could oversee several machining operations, resulted in a quantum leap in productivity.

Another giant step forward in manufacturing productivity resulted when multiple-unit, multiple-spindle machines serviced by a single workstation were conceived. This particular innovation permitted several sets of operations to be performed on different workpiece surfaces all on one machine monitored by a single worker. Through this innovation, the number of different machine tools required on a manufacturing production line could be greatly reduced; therefore, output per labor-hour jumped higher, despite the fact that the overall loading, machining, and unloading time was about the same. However, since fewer machines and fewer machine operators were required, productivity—i.e., output per labor-hour—increased.

Another exciting step forward came with the development of the indexing head, wherein a machine-tool attachment could rotate the workpiece through any required angle so that faces could be machined, holes drilled, and so forth, in definite angular relationships. In-line indexing, or transfer, made it possible for one operator to control the work being performed at several machining workstations. With this feature, the operator was capable of loading and unloading at the load station while the machine tool was still running and cutting metal. The total cycle time was reduced because the loading and unloading activity was taking place simultaneously with the machining operation.

Automation as applied to machine tools

There have been a number of developments that have made manufacturing automation possible. The programmable logistic controller (PLC), the computer control of machine tools such as computer numerical control (CNC) or direct numerical control (DNC), and, in some areas, machine vision and industrial robotics have provided the impetus for modernization of the manufacturing plant. When these facets of automation are assembled in various combinations, the basic elements (of automation) lack only the means of transportation to integrate a system.

The implications for the materials manager are fairly clear because he or she has at hand driverless tractors that can be guided by radio frequency (RF) circuits or wires in the floor of the plant, which emit RF signals. Also, conveyors of various types can convey piece parts and components from machine to machine or into several departments. Circular silos can be used to store parts between operations, or to feed them continuously (or intermittently as the case may be) to other machines or operations.

Although there are a number of ways that workpieces can be transported between machine tools, it is interesting to follow the development of the large-scale automatic workpiece indexing and linear-transfer mechanisms, which resulted in the multiple-station rotary indexing machines and the multiple-station, in-line indexing, or "transfer" machines. Machines that were originally developed as special-purpose machine tools for production, inspection, and assembly operations have now been integrated into automated machining systems. They have been called *flexible manufacturing systems* because, within the conveyorized and palletized system, pallets carrying workpieces can be shunted by computer control from one machine to another for various machining and inspection operations.

Although transportation of parts (or workpieces) between machine tools is either palletized or nonpalletized in a flexible manufacturing system, thus eliminating any materials handling being performed by an operator, the question of bringing the piece into, or to, the system is where the input from the materials management department is necessary. Let us digress for a moment. A *palletized* workpiece is usually a nonrotating part, such as a gear case or housing. A *nonpalletized* piece is one that can be picked up by a mechanical manipulator, or robot, and located (loaded) precisely into a fixture, or chuck. Most rotational parts can be handled in this fashion and it is not uncommon for nonrotational parts to also be handled in this manner by robots.

The fully automated factory

When all of the innovations in manufacturing practice are viewed from a comprehensive perspective, there is little doubt that we are fast approaching the realization of the fully automated factory. Whether or not this will be achieved in the next decade will give materials management personnel something to ponder, while there will be even more concern exhibited by much of the manufacturing cadre in the largest companies, who will be, no doubt, the first ones to pursue this goal, since all large-scale automation requires vast amounts of capital. Fortunately, or unfortunately, depending upon one's viewpoint, the smaller companies will not have the monetary resources to embark on any such futuristic endeavors and will still be plagued with all of the problems of materials movement that accompany the less than fully automatic mechanism systems espoused by the larger companies.

There will be, of course, exceptions to this generalization, an example of which is the almost entirely automatic production unit of the Allen-Bradley Milwaukee facility, which makes motor starters. It is probably one of the most advanced state-of-the-art production divisions in the electrical controls industry. The installation is well worth studying because its control and communication technology can be applied across the board in a wide diversity of production applications, from automobiles to computer fabrication.

In this facility, which can produce many different models and in any quantity of each type, one of the important considerations was to reduce the cost of materials handling, a noteworthy goal for materials management. The main goal was to remove all direct labor man-hours and to achieve flexibility in the product mix. In this regard, production and assembly operations were designed so as not to disrupt the manufacturing processes when changing from one model to the next. Another criterion of the design concept was the requirement to improve quality control while producing the units with an exceptionally fast receipt-of-order-to-shipment time.

Thus, using the computer-integrated manufacturing concept, this stockless, almost completely automatic factory can receive orders one day and ship them out the next day as the units are manufactured, assembled, tested, and packaged. In addition, even special orders can be handled on the same production line in the same way, since modifications of the models can be inserted into the production sequencing, where they are manufactured and sorted together with the more standard units. This feat is accomplished by means of a master controller, which instructs each machine to change over automatically to produce the new model and then back again to the standard model.

The whole concept of this production line is based upon the "lot size of one" thesis, thereby eliminating setup and changeover time. An important, added benefit of this automated manufacturing facility is that, because of the tightly controlled conditions of processing, a high level of product quality can be sustained. The system uses statistical quality control methods to maintain tolerances, and any unit outside of the prescribed tolerances is automatically rejected, thus ensuring uniform product quality. Since each component is tested (automatically) at 100 percent inspection levels within each manufacturing process, there is a QC cost saving, because no labor-hours are required. Testing is accomplished progressively as each part moves from one manufacturing station to the next. All in all, there are a total of 350 test and data collection points in the system.

Any large-scale attempts at fully automatic operation of a factory are somewhat remote at this writing—i.e., possible but improbable because of a number of factors, such as the complexity of some products that would require tremendous initial infusions of capital and the cheapness of off-shore competitive products due to low labor costs, which far outweigh the high cost of money in the United States due to the current interest rates. However, the realization of factory automation on the scale of the Allen-Bradley example cited and smaller applications of flexible manufacturing systems will continue to offer opportunities for the materials manager to interface with these computer-integrated operations.

Some of the ways in which materials management can service the automated production areas would be to use powered and free conveyor networks both overhead and on the ground—both tractor trains and automatic guided vehicle systems (AGVSs), and the forklift truck. Prior to the decision to use any of these transportation means, however, it would be extremely advantageous if the materials management group would devise containers, pallets, and product-carrying devices that would permit the transfer of the product into the automated system with the least amount of manual handling. If at all possible, these devices should be designed to be used directly in the automated handling of the flexible manufacturing system, possibly with an automatic transfer mechanism that will unload the product into the first processing stage of the mechanism. On the output side of the FMS, the offloading also should be accomplished automatically so that the finished product becomes the responsibility of the materials-handling personnel.

Another aspect of FMS that links it to materials management is the use of automatic storage and retrieval systems (AS/RS), which may serve as a buffer for the automated system. Parts can be stored in the AS/RS before and after processing. Conveyors or AGVSs can be used to transport parts to the FMS or remove finished components and place them into the AS/RS storage system to await further processing or outshipment.

8

Flexible assembly and materials management

Introduction

Automation in both the manufacturing and service industries has greatly reduced costs, but not to the extent that is commonly believed. Although these industrial segments might give the impression that they are low-cost enterprises, they still have costs. It might be possible, by automating production processes, to virtually eliminate direct labor-hours but almost all of the other costs remain, such as:

- Purchased materials
- Administrative and selling costs
- Engineering
- Interest
- Return on invested capital

By and large, some of these costs might be much higher than in a nonautomated plant, simply because the machinery and production equipment might be more costly to maintain and require higher labor skills, ergo, more expensive labor. Also, more complex and costly equipment requires a larger initial investment, which translates into higher interest payments.

Although the Allen-Bradley motor-starter division plant in Milwaukee, which was described in chapter 7, is very close to being fully automatic—i.e., without workers—it still requires some personnel to oversee some of the more crucial operations. As an example, a three-light warning system is used to alert the operators. A blue light indicates that a parts feeder is running low and requires replenishing. A yellow light signifies that a part is jammed up somewhere in the mechanism, or that there is a malfunction. A red light is evidence that a machine malfunction has shut down the entire system automatically. Strategically located operator displays give personnel an indication/readout and diagnostic of the condition requiring adjustment. Thus it is that even the most sophisticated of present-day automated systems in the electrical equipment, electronics, and automotive industries require human operator intervention, however small their input might be in the overall process. Also, although automation radically reduces labor costs and total costs, there are other, not so obvious costs that must be considered.

Costs of automation

One of the by-products of automation is the question of overhead costs. Although there is less indirect labor in the automated plant, the labor that must be provided to maintain the new and more complex equipment will be more highly skilled and higher priced. In addition, the greater capital investment in automated equipment raises depreciation charges and, most generally speaking, increases maintenance and operating costs. As a result of this mix, new ratios emerge; for example, fixed charges (which include depreciation) increase, while variable charges, in the way of indirect labor, decrease, making costs more rigid.

Since automation requires a larger investment in facilities, profit requirements are greater. Translating this into realistic terms means that profits on each unit produced must be higher in order to yield the same return on investment. Many entrepreneurs, however, take the position that because risks are higher with automated plants, since the initial investment is higher, automated facilities should earn an even greater return on investment (R.O.I.) than nonautomated plants. (In reality, this rarely happens because competition usually acts to dampen this wish.) In an unautomated plant if there is a business downturn and sales fall off, the management simply lays off workers in order to hold losses to a minimum. However, in an automated plant the management is at an impasse because there are few employees to lay off and it has an extraordinarily large investment in equipment that cannot be used for anything except to make a product that no one will purchase.

Another by-product of automation that should be of concern to the materials management department is the requirement for tighter specifications in the quality of purchased parts and materials. Slight deviations in quality for parts and materials usually can be adjusted readily by the workers in a nonautomated plant—a wire of a resistor can be straightened, a plate can be bent to fit, or a bracket can be forced into position. The opposite is true in an automated factory, where automated machinery might not be capable of compensating for these minor aberrations. The machinery cannot be made to stop or slow down until the part is adjusted since the automated equipment will never be as flexible as human hands, or as skilled.

The increase in standards for the parts will inevitably lead to higher purchasing costs because improvements in quality are not cheap. Inspection and quality control costs also might be increased, together with the occasional supply failure that will result when purchased parts are rejected outright or placed on hold.

The results of supplier noncompliance are usually not as disastrous for a nonautomated plant as for an automated one because if supplier failure occurs in a nonautomated plant the workers in that division are simply laid off or are elsewhere in the plant;

this procedure is costly, it is true, but fortunately, is less costly than in automated plants, where fewer employees can be released because less can be saved by layoffs, and the large investment in equipment keeps depreciating whether or not it is used to produce anything.

In automated and nonautomated plants alike, material and direct-labor costs can be eliminated, by and large, viz. materials are not consumed and workers are laid off (not without some costs, to be sure, because there are such things as supplementary workers compensation, or state workmen's compensation taxes, etc., which might have to be paid). A supply failure, even in the nonautomated plant, can be costly primarily because overhead costs continue almost unabated.

For example, if a product has a net overhead cost of 80 cents when made in a nonautomated plant, it could easily be that the fixed or nonvariable cost is 40 cents, which is incurred without regard to whether or not the product is being fabricated. Supposing the company makes a 20-cent-per-unit profit on the product. When the plant is shut down it would lose this margin of profit, thus the total loss per unit would amount to 60 cents. Therefore, if the plant is geared to produce 5,000 units per hour, then the supplier's failure that caused the shutdown would cost $3,000 every hour.

In comparison, the cost of shutdown in an automated plant is invariably higher than in a nonautomated plant. An illustrative example such as the following may explain the reason.

Let us suppose that the automated plant's overhead costs are slightly lower than the nonautomated plant's, say 70 cents per unit. A good deal more of it, however—55 cents—is devoted to fixed and nonvariable expenses, which are incurred whether the plant is operating or not. The variable indirect-labor cost that is usually expended for quality control inspectors, materials-handling personnel, and the like is replaced by nonvariable maintenance costs and depreciation expense required by the added investment in plant and equipment. Taken together, the automated plant has 55 cents of fixed and nonvariable expense compared to the nonautomated plant's 40 cents, or a difference of 15 cents per unit cost. Also, the loss of profit when the automated plant shuts down is 35 cents per unit instead of 20 cents, or 15 cents per unit greater. When the loss from per-unit profit is added to the additional fixed and nonvariable cost, the total per-hour loss for the plant amounts to $4,500 per hour based upon a production of 5,000 units per hour, or $1,500 per hour more than the nonautomated plant.

The extremely high cost of shutting down a company that contains a great deal of automated equipment sometimes works in favor of the union personnel (where there is a union), who operate this type of plant because it enhances their bargaining position. From the employee's viewpoint the company stands to lose many thousands of dollars a day when the plant is forced to remain idle. Therefore, each new strike invariably makes it possible for the worker to accrue gains in pay and fringe benefits. Automation in the manufacturing and service industries vastly increases the cost of supply failures and, no doubt, tends to increase the possibility of these happenings, and the company's top management requires that the materials management department develop the professionalism and expertise to minimize the effect of supply failures in manufacturing operations.

The flexible assembly interface

In flexible assembly systems, parts and components can be transported selectively to any of several workstations. There are several advantages of this type of installation, as follows:

- Inventories can be reduced
- Work-in-process is reduced
- Floor space is used effectively
- Assembly operations can be relocated and changed readily

The proliferation of flexible manufacturing systems (FIG. 8-1) is a result, not only of the many production and cost advantages this concept has to offer, but also of the continuing decrease in the cost of electronics and controls used to drive the systems.

As in flexible manufacturing systems, flexible assembly conveyors are being equipped with their own on-board computer and control systems. These systems have been designed with a degree of sophistication that permits the user to make changes in both hardware and software, even after the conveyor manufacturer is no longer involved in the installation.

There are a number of different types of conveyor systems that can be used in flexible assembly operations, including the following:

- Inverted powered and free conveyors
- Towlines
- Car-on-track
- Transporter conveyors

These conveyors might not look similar, but they do have certain features in common; for instance: they are modular in construction, they can be operated with a number of different electronic control systems, and they can be designed to interface with automated workstations.

Inverted powered and free conveyors

Of interest to both manufacturing engineers and materials managers are the inverted powered and free conveyors that are, and can be used as, integral parts of flexible manufacturing and assembly (FIG. 8-2). The IPF conveyor system is so constituted that free access from all sides can be attained, making all manner of assembly operations relatively easy, and the stability of the assembly platform, which is the pallet supporting the assembly fixture, can be made to withstand considerable external force applied to the product it carries. IPF conveyors can be integrated readily, also, with robotics in assembly operations, since the reach and grasp (or other manipulatory methods) of the robot remain unrestricted.

The inverted powered and free conveyor is closely related to the overhead conveyor of the same type, the exception being that instead of being suspended from the overhead structure of the plant, it is supported by the ground. This single difference means

(Courtesy S.I. Handling Systems, Inc., Easton, PA.)

Fig. 8-1. Car-on-track system used in a flexible assembly operation. Note the manner in which assemblies and components can be queued, awaiting access to each workstation. Progressive assembly operations can be carried out sequentially or selectively, depending upon the assembly program schedule.

that the engineer installing the conveyor no longer has to worry about whether or not the overhead structure can support the load, since there is hardly any weight limitation imposed by most factory floors. Other advantages are the ease of maintenance, the absence of drip pans to catch oil and debris, and the fact that the conveyor carriers are work tables, where assembly operations can be performed without regard to the influence of gravity.

Towlines

In-floor, drag-chain conveyors have long been used in the automotive and heavy-industry fabrication plants to form progressive assembly production operations. Again, hardly any weight restrictions are imposed on these conveyors and, since running

(Courtesy of The Jervis B. Webb Co., Farmington Hills, MI.)

Fig. 8-2. An inverted powered and free conveyor used in flexible assembly operations.

speeds can be varied for the line as a whole, assembly operations and combined operations can be programmed to suit whatever speeds are desirable. This type of conveyor usually paces the operation, and workstations can be established anywhere along the path of the dragchain. Also, since air-rights are not sacrificed with these installations any workstation can be serviced by overhead conveyors, hold conveyors, and the like.

The in-floor drag-chain conveyor is very safe and relatively quiet in its operation. The one obvious disadvantage is the question of cost, since in-floor conveyors require considerable outlays of capital. Another disadvantage is that once this type of conveyor is installed, modification or removal is inordinately expensive.

Car-on-track conveyors

Car-on-track (COT) conveyor systems, like inverted powered and free, towlines, and transporters, are relatively expensive installations. Transporters are probably the least expensive in this group of conveyors.

The original car-on-track conveyor was designed and developed by S.I. Handling Co. of Easton, Pa., a leading company in automotion engineering. Their version of car-on-track is a patented system called Car-Trac. This COT conveyor has carriers that ride on two rails with a drive wheel between them for propulsion. The drive wheel underneath each carrier provides contact with a drive tube so that the forces of acceleration and deceleration can be controlled minutely and precisely by varying the contact angle between the drive wheel and tube. This car-on-track system is extremely well suited to flexible manufacturing and/or assembly inasmuch as routing flexibility can be accomplished easily by transferring cars from one track to any other.

Some of the S.I. Handling systems have been designed to handle and maintain a position precision of 0.005 inch in three axes. It is obvious that this feature is immensely desirable where interfaces with robotic or other automatic assembly workstations are necessary and where the repetition (repeatability) of positioning is crucial. Naturally, this degree of positioning costs money, and the higher the precision the more expensive such an installation will become. Nevertheless, cost trade-offs can be considered when the choice is between expensive robotic or machine vision systems required to adjust to the carriers "at rest" position or the degree of precision required by the car-on-track conveyor. Other considerations, of course, would be the number of robots, work stations, amount of maintenance required to maintain the degree of precision, and so forth.

Transporters

Of significance to the material manager is the use of transporters to service the requirements of flexible assembly operations. Various forms and degrees of flexible manufacturing and flexible assembly operations can be assumed to be interchangeable. It is only in the degree of position repeatability that the two concepts vary, and even this isn't a reliable gauge of interchangeability between the two concepts, since a high degree of positioning might be required in either manufacturing or assembly.

Transporters are combinations of roller, chain, and belt conveyors that are used to convey tote boxes, trays, and specially designed carriers of many configurations in and around the geography of the manufacturing plant. The carriers can be fed continuously or intermittently from overhead conveyors, which dump or place their loads on carriers, or the beginning of the assembly operation might be initiated by a worker loading the transportation carrier. It is possible, in some instances, to program robots or mechanical manipulators to perform this task when the operation is highly repetitive.

9

Materials handling
as an element of
materials management

Introduction

In the overall philosophy of materials management, I have attempted to demonstrate that one of the most important elements of this concept is the flow, or movement, of materials from the supplier through the manufacturing and distribution channels and, ultimately, to the consumer. In order to accomplish the most effective handling and least total cost of the movement of materials through the system, serious consideration must be given to the means by which this movement will take place. It is not sufficient to say that we will handle material using—what else?—materials handlers because the "systems concept" militates against the possibility of operating in a vacuum. The very essence of *"system"* means that every bit and parcel of the manufacturing entity bears some relationship to each and every other bit, and what is of concern to one plant department is also of concern to another.

This is the very reason for materials management in the first place—the business of materials management is to tie together all of the loose bits and pieces of materials movement. That is why, in the optimum organizational structure for materials management, the organizational entity should maintain control of the purchasing function. As I have indicated, purchasing is concerned not only with quantity, price, and quality, but a number of other specifications that influence the flow of materials. Examples are the way in which material is packaged (or not packaged) by the supplier, i.e., the types of inner, intermediate, and outer packaging of materials, parts, components, and assemblies; the shipping schedules, lead times, and carriers to be used; the way the parts are to be shipped whether in bulk or separately; the protective packaging to be used or the absence of packaging; and the protective coatings to be applied, when required. These topics and a range of other considerations on the subject of packaging are discussed in the next chapter.

Materials handling methods

Materials handling is the physical side of materials management and has the responsibility of assisting in the movement of materials within the plant. Since the fully automated factory is still some distance in the future, we do not have, at present, an assemblage of various types of computers and automatic machines that operate in a workerless void.

At this period of our economic society, we find that the degree of automation varies considerably from plant to plant and from industry to industry. Nevertheless, the well-integrated manufacturing complex is, and should be, designed around its materials-handling system. The materials handling between machines and departments must provide for the free flow and passage of materials, personnel, and equipment, without blockages or cluttered aisles.

In flexible manufacturing, machine tool builders, both in the United States and abroad, have the capability of producing complete, automated machining centers, in which the machine tools automatically change their own tools as wear occurs and tolerance limits are approached, built-in diagnostics enable the machine to do their own troubleshooting, and the materials handling is performed automatically. It is also possible, at this time, to go a step further than machining and bring the parts into an assembly and test area with completely automatic operations. Yet, although all these things are currently possible, and are being done, the vital links between the automated systems and the rest of the factory world lie with the more conventional methodologies of materials handling. Forklift trucks, tow conveyors, drag-chains, and all types of overhead and gravity and powered conveyors are still being used in the manufacturing plant and the distribution areas.

Container requirements

The materials-handling engineers of the plant should work closely with purchasing, manufacturing, and quality control to design containers that fit the parts, or families of parts, to be handled. Several things are accomplished as a result. The well-designed container will help to maintain the quality and integrity of parts, especially where machine finishes are crucial. Also, there would no longer be a need to dispose of packaging and disposable dunnage (although reusable dunnage might be an effective trash eliminator; however, this means returning dunnage to the point of origin).

Although the engineered, or customized, container might have a higher initial cost in contrast to an off-the-shelf type, if the frequency of use is high it will tend to pay for itself in the long run. However, where critical machine finishes are concerned, even a high initial cost might be justified in terms of the elimination of scrap or reworking of the damaged part. In many instances, rework of parts might be impossible and the part that has had several hours of expensive machine tool or labor-hours might pay for several, fairly expensive, customized containers.

Other benefits are that housekeeping is better, plant neatness is a morale booster, and fire hazards and safety problems are virtually eliminated.

Using properly designed containers that fit the handling methods and equipment will tend to eliminate much of the waste that is inherent in the use of corrugated paper containers, wood boxes, crates, and pallets, with their accompanying combustible interior packaging. Good substitutes for these packaging materials are stackable, collapsible steel pallet boxes, stackable plastic racks, pallet tray packs, steel or plastic bins, and other types of returnable containers. In addition to being well suited to the production and assembly lines, these functional containers can be designed so that they will fit modularly (ergo, economically) within motor and rail freight carriers, and in-plant and interplant transportation equipment.

Since a plant's investment in captive containers might be sizable—i.e., from a few thousands of dollars in small plants to several millions of dollars in larger plants (one plant I was involved with had over $60 million invested in containers—it is advisable to maintain accountability for the number of containers shipped to and received from suppliers. Steel containers can have numbers embossed in convenient and visible locations, but other means are available. An embossed metal tag can be attached to the handle or latching device of the container and the latest automatic identification means, either bar coding or RF tags, can be used. If the latter methods are employed, an added benefit would be the ability to track the container using in-plant computer terminals on the shop floor and in the receiving, shipping, and warehouse areas.

Receiving and shipping

Areas used for receiving and shipping operations often seem like afterthoughts, instead of being designed to handle incoming and outgoing shipments efficiently. A brief summation of how receiving materials handling can be improved will follow from an examination of the area. It can be observed by random sampling if hand-unloading of incoming receivable happens often. If this seems to be the case, then changes in packaging, palletizing, or containerization might be in order.

Are receiving personnel using the proper equipment to unload carriers? If they are not, then it would seem feasible, if the receiving operation is to be improved, that these employees be provided with the proper tools and equipment to perform at their highest levels.

Another consideration is the space available for receiving operations. If not enough space has been allocated for this function then it behooves management to rectify the condition. If there is more than one receiving area in the same plant building, it would make sense only if some form of Just-in-Time manufacturing were being practiced, since JIT requires frequent and closely timed delivery of receivables at the closest point of use. Another point is that either receiving or shipping docks should be centrally located to all points of use. That is one good reason for locating these two departments contiguously.

It is also possible to increase the speed and effectiveness of truck turnaround time—i.e., the portion of time that the tractor-trailer sits at the dock waiting to be either loaded or unloaded—by having a fleet of captive vehicular (tractor-trailers) equipment equipped with a slug loading and unloading system. In this system, powered roller conveyors bolted to the bed of the trailer are capable of disgorging the entire contents of the trailer in a matter of minutes, whereby the entire contents are propelled onto dockside conveyors. The shipping department would have similar equipment to effect slug loading.

If, during your random observations of either shipping or receiving, you happen to see idle employees on not more than one or two isolated occasions, you can rest assured that the departments are being staffed properly for their responsibilities. Persistent idleness on these docks indicates shoddy supervision, and steps should be taken to correct the situation.

Some physical aspects of receiving and shipping concern the adequacy of truck dock spots, about 12 feet center to center for adjacent spots; the availability of dock plates or levelers; and doors of adequate height and width. In northern climates, shrouds should be used to keep cold air out of the building when the doors are open to receive the truck at the working face, and dock lights should be provided, at the small cost of $50 or $60 per extension spotlight.

After these basic needs are taken care of, one of the major problems that can plague both receiving and shipping has to do with congested space, which can be a result of faulty housekeeping. More often than not, however, either the receiving system is not working properly or there is simply not enough space to handle incoming materials in an adequate manner. Dunnage and packaging materials disposal requires a systematized method for handling trash.

Quality control and inspection are functions that should be closely integrated with receiving. Cartons and sometimes containers that are opened will have parts and components removed for testing. The standard operating procedure should be to indicate that there is not a full count therein; otherwise another inventory might show an incorrect quantity. Also, merchandise that is being held for quality defects until such time as adjustments are made should be so indicated on the container in a formal manner so that everyone in this or related departments is aware that the merchandise is awaiting adjudication, either for rework, replacement, or new price considerations. In other words, a discount might be in order because the piece is 0.25 inch too short, but it might still be used on some modes, etc. Lastly, there is the question of packing lists and the checking of materials. Both of these subjects should be investigated to assure the materials manager that both are adequate.

By and large, this discussion has focused on receiving functions, but the two areas have so many parallel functional responsibilities that the two operations are quite similar, with only a few exceptions—one being that the shipping functional area should be located as close to the final operation of fabrication as possible, for obvious reasons. If there is the necessity for applying protective coatings, painting of parts, and packaging—and this is a general thing for all products—then the shipping department should be located contiguous to the last operation prior to being loaded out. Also, blocking and bracing, when required in the normal course of business, should be placed close to the shipping department. Sometimes blocking and bracing materials can be obtained from a subcontractor, or even better, from a nearby, sheltered workshop.

Linear vs. U-form materials handling and JIT manufacturing

Materials managers who desire to improve materials handling and reduce inventories have looked hard and fast at Just-in-Time (JIT) manufacturing techniques, and what they have found is that not only does JIT accomplish these objectives, but there are additional benefits, as well. Because of the strict discipline associated with this concept, suppliers, as well as factory workers, perform better, quality levels increase, and waste is minimized. JIT virtually eliminates excessive inventories because it places the scheduling burden directly on suppliers.

There are a few drawbacks to this aspect of JIT, one of which is that when a supplier's plant is faced with a potential strike, it is often necessary to beef up inventories in what is known as a *strike hedge*. Strike-inventory reserves waste storage space, tie up capital, and have other disadvantages normally associated with an overabundance of supplies. Another drawback is the amount of discipline that must be exercised by the manufacturer and exerted on the company's suppliers in order to keep schedules finely tuned. When JIT is working well, however, it more than makes up for any disadvantages that might surface. Thus, every manufacturer should strive to use some small part, if not all, of this concept because the benefits are huge.

JIT manufacturing, as the name implies, is a methodology that makes the supplier responsible for quality assurance and supply of a prescribed quantity of materials to the plant's production lines at established time intervals. When JIT techniques are to be installed, it will be necessary to increase preventive maintenance schedules for all machine tools in order to increase the utilization percentage. In addition, machines might have to be rearranged to straighten out the flow of materials through the shops. Since parts availability is increased with the JIT program, only the required parts are machined and there are fewer work-in-process (WIP) banks. Also, cutting machine setup time will increase machine availability as an added scheduling benefit.

JIT systems installed in the United States have significantly greater problems in implementation than they would have in Japan, the JIT country of origin. A few of the roadblocks that must be surmounted are our proverbially distasteful labor-management relations. The attitude that "they're agin us!" is difficult to overcome, and since teamwork by all layers in the company is essential in these programs, careful orientation and implementation are required. Warehousing of parts and inventories might be eliminated in some programs, but normally there is a reluctance to abandon safety stocks, especially in the light of the strike hedge philosophy just discussed. As an example, a plant might receive material shortly before it is ready to use it; however, if you have a continuous throughput, you should have little or no requirement for warehousing. Also, if parts cannot remain outside a factory building because of space limitations or inclement weather, it might still be necessary to maintain a warehouse facility. Thus, the warehouse would be used as a buffer adjacent to the manufacturing production areas.

Another variation of JIT, which requires a good deal of supplier confidence, provides that the company receive the parts but does not actually pay for them until they are consumed in the manufacturing process. In this fashion, the company can take advantage of the accounting practice of having a float, just as most of the automotive companies have done for years. This stratagem might not contribute to decreased inventories, nor permit rapid engineering changes to be made; however, defective parts, which have not yet been paid for, can be kept separate in the warehouse and be returned without tying up capital.

Since plant manufacturing space is very costly—surprisingly, a square foot of plant manufacturing space might cost even more than a square foot of luxury-level residential to use a subsidiary warehouse to support the operation. The buffer warehouse could be company-owned, or it could be owned by the parts supplier, especially if a long-term relationship has been enjoyed by both companies.

In the automotive and electronics industries, parts plants are being located in close proximity to the assembly divisions of manufacturing plants, thus fulfilling some of the requirements of the JIT methodology. In some portions of the electronic industry segment, the parts fabricated by suppliers are delivered to the company and placed in mini-load (small AS/RSs) or carousel storage units until all of the components are on hand. When all parts and components have been delivered, the materials are then delivered to the production lines, where they are assembled to fill custom orders.

Naturally, having suppliers consign materials for a manufacturer will work only where there is tremendous purchasing power, as in the case of the large automotive and electronics manufacturers. In the main, most parts suppliers do not have the cash reserves required to finance their manufacturing associates, especially when they are forced to relinquish control over their own products.

In JIT manufacturing, parts can be received on a weekly, or even an hourly, basis, depending on the size of the parts involved. JIT eliminates bulk storage areas adjacent to assembly lines in the plant. The only storage requirements that are necessary are for long-lead-time items, although even this requirement could be eliminated if the supplier's location were in proximity to the plant.

JIT plants usually rely quite heavily on the use of overhead powered and free conveyors, upon which material is stored until it is required at the point of use on the assembly line. From the materials handling standpoint, storing parts on overhead, accumulation loops (gravity or powered) is a good way to use air rights in a plant in order to release valuable floor space for production operations.

Also, in JIT programs receiving points are spread throughout the plant so that suppliers can place their merchandise in the closest possible place to the point of use. This is another value change requiring discipline on the part of plant employees, since the decentralized receiving function is much harder to control than a centralized receiving area would be.

In chapter 6, MRP programs and procedures were discussed, and you were advised concerning MRPI, the forerunner of MRPII. The former concept was called Material Requirements Planning, and the latter was named Materials Resources Planning, MRPII being the more comprehensive of the two plans.

Although MRPI is quite capable of controlling inventory and the flow of material through the plant, a JIT program is vastly superior. However, MRPII has exhibited better control and more widespread plant involvement than MRPI; thus it has some advantage over JIT in a U.S. plant environment, although there are some major differences in ethnicity, labor-management relations, and so forth that militate against a widespread or thorough application of JIT as practiced in Japan. Nevertheless, when a U.S. manufacturing plant has been conformed to accommodate JIT, inventory or stock turns might jump from 3 or 4 to over 12 turns a year. Lead times, normally measured in months, might decrease to days, and software programs require periodic adjustments to help them keep pace with the growth in productivity.

The process sheets used by production planners habitually indicate all of the actions and functions used to fabricate a piece part. If their analyses are very detailed, and it is not unusual for them to be so, the time for each action, distance traveled, quantity and kinds of tools used, and similar information will find its way into the process sheets.

For example, when speaking of distance traveled, the production planner could be indicating the number of feet (meters) that the part or forklift truck must travel to carry the part through the production process. The planner has to keep asking himself questions, such as: what is the necessity for doing a particular operation? Is there a better way of doing this? and so forth. When manufacturing engineers review the process sheets, they might decide to rearrange the tool lines so they are closer to the machines that are using the tools or substitute machine tools different than the ones selected by the processor.

On the other hand, when using JIT methods, the chief objective is to reduce the elapsed time between operations. In this philosophy, if one operation is at the end of one production line and the sequential operation is at the other end of the line (as an extreme, but not uncommon, example), then JIT performance suffers. Thus, the conventional plant with only one receiving dock is at a serious disadvantage, especially when, compounding this inconvenience, the storage area and tool cribs are traditionally located in one centralized area and the machinery departments are arranged by machine types.

Transforming the plant to take advantage of the best that JIT has to offer means that storage points have to be located close to the production operations; expendable and other tooling must be conveniently available to the machines, even if pneumatic tubes have to be used in order to achieve this objective; and more than one plant receiving point will make it possible for parts to arrive at their points of use in such fashion that production and assembly schedules can be achieved on a timely basis.

Although the JIT concept appears formidable from the overall view of a plant that is completely immersed in this philosophy, one of the commendable factors concerning the JIT method is that the plant metamorphosis can be accomplished in small, bite-sized pieces. Thus, one phase of slanting the conventional plant to JIT is the rearrangement of machine tools from the customary straight-line production tool array to the cellular U-form, or sepentine, design of layout. The Japanese call it *mizusumashi* and it is contrary to customary U.S. manufacturing engineering practice for the general run of machine shops. However, if you step back to view the overall manufacturing scene, you can see the cellular arrangement appearing in FMSs, where machine tools are usually arranged in a closed loop with a miscellany of different machines. Therefore, we could just as easily have called FMS cellular machining as not.

The same statement could be made for flexible assembly; thus, it is a matter of scale we are discussing at this point. It is just that the application of JIT can be made by using our conventional machine tools and simply rearranging them in the *mizusumashi* fashion. The one difference in this simplistic analysis is that in FMS we still have to bring the tooling, storage, and parts to the site. In JIT we must rearrange the site to accommodate the peripheral necessities.

In continuing the discussion, in conventional factories, materials handling is performed by forklift trucks that pick up containers, parts, unitized loads, and the like and travel down the production line to the next machine in the processing sequence, or the forklift truck wends its way through a convoluted maze from one type of specialized property—a company that has a high rate-of-flow process or production line might want machine tool to another, and sometimes to a machine tool, or process, that just happens to be in another department separated by a fire wall. More often than not, the forklift truck will return empty (which is almost sacrilegious to an industrial engineer) to start the movement all over again. This repetitious task is performed many times over, sometimes for a full working shift.

Although this example represents some of the "unthinking" materials handling performed in many plants, it is such a wasteful practice that it is easy to recognize the statistic that many piece parts are worked on (that is, in processing or machining) for only 2 percent of the time that they are in the plant. The other 98 percent of the time is expended in transportation and storage, with the preponderance of the time spent in storage, usually awaiting transportation, or queued up waiting for a machine.

With JIT materials handling, machine tools are generally arranged in U-shaped cell, and all of the cells in a department can be arranged in a much larger U-shaped configuration. Therefore, under this discipline, forklift trucks travel relatively short distances and their productive time is increased in large measure.

10

Making materials management work

The necessity for standardization

There are three possibilities, or conditions, that direct companies' efforts toward standardization, particularly in the area of packaging of incoming parts and components. In the first place, if a company has embarked upon a mechanization program then the need for standardization is fairly urgent, since package size must be compatible with the mobile handling equipment and storage mechanisms that will be employed. Highrise, stacker crane, and other automated storage and retrieval systems require a rigid adherence to certain restrictive package sizes. These automated systems require certain standardization of packaging and load sizes in order to receive and store materials in the most efficient and economical manner possible without requiring that incoming (receivables) be repackaged on arrival at the storage site.

The full effectiveness of the automated systems can be achieved only by a concerted effort on the part of suppliers, purchasing, and materials handling in order to resolve packaging and handling problems. Whenever there is any doubt regarding particular packaging requirements, or when a supplier has difficulty in complying with the company's packaging specifications, the supplier should direct its inquiries to the appropriate buyer in the company's purchasing department. The supplier should rigorously observe this chain of command; otherwise complete chaos will result.

In the second place, if a company is contemplating a mechanization and modernization program then, for the reasons previously given it would be feasible to parallel packaging standardization by alerting its suppliers to the necessity for considering and implementing the new requirements.

Finally, even if a company is not mechanized, nor is contemplating any form of modernization in its handling and storage techniques, a certain uniformity in packaging sizes, weight restrictions, and materials might be necessary in order to store and handle incoming merchandise as efficiently as possible. Consider these examples: a large container of stampings, weighing in excess in 4,500 pounds, when the largest available forklift truck in the plant has only a 2,500-pound capacity, and an incoming cubical, corrugated container measuring $5 \times 5 \times 5$ feet, when the bulk storage rack spacing has a depth of 48 inches and a spacing height between shelf beams of only 50 inches. Surely there is a need in these instances for packaging specifications that will place the onus of compliance on the supplier, and eliminate the costly labor and confusion these vexing problems create in the receiving department.

Most of the time, the use of packaging specifications do not add a particle to the cost of the product. Almost every reputable supplier will do what it has to do to satisfy its customer and stay ahead of the competition. There is only one company that I am familiar with that is adamant about changing packaging standards to suit the customer, and that is because its method of manufacturing and the proprietary product involved have virtually eliminated all the competition in the field.

Every company, including even the smallest firms, once they have graduated from the alley-shop category should consider the benefits of having a set of codified instructions to govern the packaging of its receivables. If annual sales are over $2 million, it pays to have a packaging manual to guide suppliers because it will result in greater efficiency in receiving and handling and will, no doubt, achieve kudos for the materials management department, as well as get the job done faster, better, and with greater precision. (See chapter 13 for a discussion of weigh-counting for inventory control.)

A packaging manual

All manufacturing plants receive materials with a certain regularity and frequency, especially as annual sales climb past the lower limit of $2 million. When this condition exists, the company needs a packaging manual in order to standardize handling methods and equipment, and to reduce the cost of operations and further increase the effectiveness of the receiving activity.

In the systems approach to materials management, since we are greatly concerned with every facet of materials movement, including materials handling, we must carefully examine every factor that impinges upon the cost and efficacy of the handling function. As an example of this intense and purposeful scrutiny, let's take a look at packaging. Poor packaging can be a result of a number of causes, such as, incorrect container size—too large a container size with overweight loads, too small a container size requiring excessive manual handling, improperly unitized or nonunitized loads, expendable wood pallets incapable of sustaining the load, deteriorated pallets with exposed nails, etc.

Taken altogether, poor packaging is unsafe, wastes time, and causes untold disruptions in the orderly flow of materials through the system. The solution, then, for this problem is to attack the problem at the source, which in almost every instance is the supplier.

Internal and external communications

Although it is true that the supplier should be the focal point of the strategy to obtain better packaging, there still remains the matter of obtaining in-plant support for the new look at packaging. If other departments in the plant have had complaints about packaging, you're off to a good start in obtaining a plantwide cooperative effort.

It is at this point that you or the materials management group bring(s) up the subject of a corporate packaging manual. To further bolster this effort, it would be advisable to obtain statistical data indicating lost labor-hours and photographic evidence of parts damage resulting from faulty packaging to support your claims.

Since the packaging manual should represent the best thinking in the company on this subject, as you develop various elements of the manual you should pass them for review and comment. Involving each concerned department will add more strength and support for the program, and the packaging project will be viewed as a cooperative effort. After all, it's a companywide concern that is being resolved.

It is worthwhile obtaining departmental cooperation—i.e., input from traffic, purchasing, receiving, quality control, warehousing, and involved manufacturing groups— since their practical experience will keep the program on a realistic footing and will tend to weed out the academic niceties that add words without substance and prevent the manual from achieving its full potential as a working tool of materials management.

Although cooperation within the company is essential to the success of the program, it is also necessary to obtain cooperation from the many suppliers who will be affected by the new company policy. It has been my experience that it is possible to obtain a list of the chief offenders (suppliers with notoriously poor packaging records) from the receiving department. These suppliers, and some of the larger companies selling to your plant, should be visited directly by members of the materials management team who are compiling the manual. Since the packaging manual will spell out the requirements for identifying the items to be supplied, the kinds of packaging, communications requirements— such as who the supplier contacts for information—and the like, the suppliers contacted should be able to provide realistic input into the formulation of the manual. Keep in mind that the manual must be a practical, working tool for both the company and the supplier.

The working environment

Whatever has been said in the last section concerning packaging requirements, does not relieve the supplier from its responsibilities for providing adequate protection, in the form of packaging, for the products that are being supplied. For example, the supplier must comply with all laws and regulations, including but not limited to:

- The railroad's "Uniform Freight Classification"
- The trucking industry's "National Motor Freight Classification"

If the company is large enough to have either a materials handling engineer, a packaging engineer, or both, then they will no doubt be involved in the formulation of the packaging manual as part of the materials management task group. Since the purchasing department, however, is and should be the first contact point for the supplier-company communication attempts, both the packaging and materials handling engineers should observe this protocol carefully, despite any prior contacts they might have had with the supplier in the course of working on the packaging manual.

In establishing contact with suppliers on the subject of the company's packaging program, it would be well to engender the feeling upon the part of the supplier that he or she is free to innovate—to make suggestions and recommendations that will help lower packaging costs while improving packaging. The packaging manual should serve as a guideline and should not be construed as a restrictive device that will hamper improvements in packaging. Thus, the packaging manual should help each supplier utilize standard packaging materials and methods to their fullest extent. Therefore, the supplier's constraints will be the minimal loading and packaging requirements indicated in the manual.

One of the constraints, if you would go so far as to commit yourself to this concept, is that the supplier should be able to indicate, prior to shipment, the manner in which the product will be packaged. To formalize this step, it is advisable to use a standard format that will provide all the information at one time. For this purpose a "Packaging and Shipping Data Sheet" should be developed (see FIG. 10-1) containing the required packaging criteria.

PACKAGING & SHIPPING DATA SHEET

Supplier Name and Address   Return to:_____

_____   _____

_____   _____

_____   _____

Date_____ Part/Model No._____ Description_____
Status: New ☐ Revised ☐ Note: For any changes to be made in Packaging or Shipping Methods, see Plant Purchasing Dept. for prior approvals.

Part or
  Model   Length   Width   Height   Weight   Materials

	☐ Metal	☐ Rubber
	☐ Glass	☐ Fiber
	☐ Plastic	☐ Other

Packaging Data

When manually handled packages are unitized with mechanically handled unit loads, complete both sections.

Manually Handled	*Mechanically Handled*	*Closures*	*Banding*
☐Loose	☐Palletized Boxes	☐Glue	☐Metallic
Bundle	Pallet Tray Pack	Tape	Nonmetallic
Bag	Pieces on Pallet	Staple	Wire
Drum	Pallet Box	Wire	Other

Corrugated Box:		Cord	*Rust Prevention*
Reg. Slotted	Corrugated-Wood	Other	Type:
Telescopic	Wire-bound	_____	(Explain)
Full-Flap	Crate		_____
	Returnable Company-Owned	*Interior*	

Fig. 10-1. An illustration of the elements of packaging and shipping data that is used by the supplier prior to shipping materials to the plant, and for which approval must be obtained from the receiving company.

Material _____ Returnable Supplier-Owned Wrapped _____
Bursting   Other Type _____ Loose _____
  Strength __ lb Send drwgs, specs, photo   Cells
Other _____   *Load Size:*   Liner
Load Size: L _____   Diecut
L _____   W _____   Nested
W _____   D _____   Other
D _____   Gross Wt. _____
Gross Wt. _____ No. of pieces _____   _____
No. of pieces __

Description of Pallet Design:
☐ 2-way   ☐ 4-way   ☐ Block   ☐ Stringer
☐ Other_____
Size L _____ W _____
Fork Entry: Width _____ (Min 22 inches)
            Height _____ (Min 3⅝ inches)

Container Charges:
☐ Supplier Returnable: Type _____
   Deposit required: $_____

☐ Expendable (one-way): Type _____
   Pallet cost: $ _____ Packaging Cost: $_____

Type of Carrier to be used:
     ☐ Rail ☐ Truck ☐ Other _____
If part or model cannot be packaged by means of existing container types: (Explain)
_____
_____
_____

Additional Comments: _____
_____
_____

Supplier's signature _____ Title _____ Date _____
For Company Use Only:
   Purchasing_____ Date_____
   Matl. Handling Engineer_____ Date_____
   Packaging_____ Date_____

(Note: An instruction sheet should accompany this form, but this can be placed on the back of the form.)

Fig. 10-1. Continued.

The information provided by the supplier must then be reviewed by the company's packaging engineer (presumably in the materials management department) in order to determine whether or not the packaging will be of sufficient durability to withstand the rough treatment of materials handling in both the supplier's plant and the receiver's plant, the hardships of in-transit shipping, and the impact forces it must withstand when unloaded/loaded at trucking transfer or consolidation terminals. This procedure also covers railroad shipments. When LTL (less than truckload) shipments are dispatched, they are usually picked up by local trucking firms and, according to the distances traveled, might be consolidated with other loads at least once prior to delivery at their final destination.

Sometimes it is not possible, by merely reviewing the Packaging and Shipping Data Sheet, to determine how well the merchandise will travel, especially if it contains some fragile components or has a relatively complex configuration. In this event, you might need to have the supplier certify that his packaging is acceptable and sufficient for the intended protection of the merchandise to be shipped.

Preshipment testing of packaged materials is a method that can be used to reveal packaging deficiencies prior to the actual shipment of merchandise. A number of approved testing methods can be used to ferret out "underpackaging," or "overpackaging" so that these defects can be corrected. There is nothing inherently wrong with overpackaging, except that it is a wasteful practice that is either passed on to the buyer, or that robs the seller of an additional margin of profit. A reputable purchaser does not want to withhold a just profit from his or her supplier, hence the view that somewhere between "not enough" and "too much" is the right amount of protective packaging that will do the job and increase the profitability of both buyer and seller.

Preshipment testing can be accomplished by any certified testing laboratory. Almost every major city has at least one such laboratory. They use standardized equipment and procedures to discover packaging deficiencies that should be corrected prior to shipment. Some of the tests used are, as follows:

- Vibration
- Drop
- Incline-impact
- Compression

In addition to these tests, there are a number of standardized materials tests that can be made for the purpose of isolating and correcting packaging problems. Figure 10-2 is an example of a form used to certify the preshipment testing of packages. It can be used by any supplier as a verification of the worthiness of his/her packaging method. This document, when properly completed, should be sent to the purchasing company's buyer prior to shipment of the merchandise.

Kinds of packaging

Receivables, or in-bound freight, by and large consist of packaged materials in corrugated cartons, palletized unit loads, and containers. We can break these categories into two broad divisions:

- *Manually handled packages* Packages that are not suitably or conveniently handled by a forklift or other industrial truck, and that usually weigh 50 pounds or less.
- *Mechanically handled packages* Unit loads that are palletized, placed on slip-sheets, are on skids, are shrink- or stretch-wrapped, containerized, and the like, in such a manner that they can be readily handled by any forklift or industrial truck.

Since the packaging manual is a means of educating the supplier to the company's requirements, you must do everything you can to explain what you want the supplier to

CERTIFICATION OF PRE-SHIPMENT TESTING

Part No. or Model No. _____ Date _____

Name or Description _____

Supplier's Name _____

Address _____

Shipped From _____

Packaging

Laboratory Name _____

Address _____

Package Design   Design of Prior Test _____ ☐ New   ☐ Revised

Description

Dimensions

Interior Packaging

No. of Parts/Container   Gross Weight _____ lb

_____ kg

COMMENTS:

_____

_____

_____

We hereby certify that the package and contents described above have passed the Shipping Container Pre-Shipment Test as prescribed in the "X" (your company) Packaging Manual.

Signature of

Authorized Supplier Official _____ Title _____

Date _____

Fig. 10-2. A form to be used for certifying package testing prior to shipment.

do when he or she is shipping us material. Therefore, you should explain in an explicit manner what you mean when you talk about *manually* handled or *mechanically handled* packages.

Manually handled packages   The following characteristics describe a manually handled package:

- Depending upon the type of part to be contained, a corrugated box (popularly called a cardboard box or carton) is the most suitable type of container for a manually handled package.

- The package must provide proper and adequate protection for the parts so that they arrive at their destination in good condition.
- The gross weight of each package should not exceed 50 pounds, and its shape should permit easy handling by one person.
- The package must be securely fashioned so that the contents will not spill out, nor should the fasteners or closure method constitute a safety hazard.
- In certain instances, package closures such as staples or other metal fasteners should not be used.
- Although glue or other adhesives can be used as package closures, it should be applied in strip or dot form so that it has sufficient adhesive strength to withstand shipping and handling, but still permit the container to be opened without difficulty.
- The packaging materials may be recycled, but the container should be an expendable, or one-way, type, unless the form of packaging to be used has been approved by both materials handling (package engineering) and purchasing.
- Parts that are a meter or so in length should be able to be removed from the end of the container.
- When using corrugated cartons, eliminate void spaces. (Most damage to packaging of this type is experienced when the voids in the containers cause stacked containers to collapse.)
- Bundles are permitted only when the part configuration or other kinds of packaging become inordinately expensive for the part.
- The use of second-hand (reused) corrugated cartons should be permitted only after obtaining approvals from the purchasing department and materials handling (or packaging engineering).
- In some instances containers might have to be tested prior to shipment. This is especially necessary for electronic components, ceramics, and other fragile parts and components. A preshipment certification might be required (see FIG. 10-2).

Mechanically handled packages Requirements for mechanically handled packages are as follows:

- A unitized load must have a strong-enough pallet base to ensure that the load arrives at its destination in a safe condition and can be handled thusly, as far as the first point of use inside the plant. It is important to establish expendable (one-way or one-use-only) pallet standards. Other types of containers and packaging materials should be expendable also, unless an understanding or approval is obtained through the purchasing department for a "returnable" container.
- Drums, barrels, bags, and salvaged containers that have been reused are more difficult and costly to handle—i.e., they require more labor input—therefore, they should be used only with prior approval from purchasing. The gross weight of palletized or unitized loads should conform to the restrictions established by the materials handling department and administered by purchasing. (The materials handling department should advise purchasing of load limit sizes based upon materials handling equipment capacities.)

- Palletized loads should have an underclearance dimension of 3⁵/8 inches and a minimum width for forklift-truck entry of 24 inches.
- Pallet loads should be stacked with cartons in an interlocking pattern and stretch-wrapped to ensure stability.
- For heavy palletized materials, the loads should be securely banded with edge corner protectors, when necessary.
- Strip gluing or stretch wrap should be used on all pallet loads to ensure stability in transit.
- Pallet overhang should be minimized or eliminated if damage to the carton contents will result.
- Corrugated board containers are preferred over wirebound or wood boxes, when practical, since corrugated containers can be recycled more readily.
- Palletized loads should consist of only one part number, whenever possible, to minimize errors, and make receiving, checking, and piece count easier.
- The company's rust-prevention and anticorrosion guidelines should be adhered to if the parts could be adversely affected during shipment and storage. (See later in this chapter on preventing corrosion.)
- The packaging manual should include definitions for various types of containers, such as:

  o pallet boxes (corrugated, wirebound, or wood)
  o pallets with loose parts
  o boxes on a pallet
  o pallet tray pack
  o pallet shrink wrap
  o pallet stretch wrap
  o crates

- Instructions should be available so that piece parts can be packaged properly. Individually wrapped parts are not desirable, although some parts might require additional protection; in that case, cellular or die-cut, corrugated egg crates might be necessary.

Fasteners and small parts   Packages containing fasteners and other small parts, such as nuts, bolts, screws, washers, and the like, are very dense, and even small quantities of these items can be very heavy. These parts should be packaged in accordance with the fastener industry's container standards, which comply with applicable carrier regulations. The following rules should apply:

- Small parts should be packaged in fiberboard boxes (fiberboard is denser than corrugated board) or in heavy, bulk-style expandable, corrugated containers. The following sizes are available:

Container style	Size (in inches)		
1/16 keg	6¹/4 ×	6¹/4 ×	3¹/8
1/8 keg	6¹/4 ×	6¹/4 ×	6¹/4
1/4 keg	9 ×	9 ×	6¹/2

Container style	Size (in inches)		
1/2 keg	9	× 9	× 13
3/4 keg	11	× 11	× 12 1/2
Full keg	11	× 11	× 17 1/2

- Dense small parts require well-constructed expendable (one-way) pallets, since the parts are dense and will weigh more than other cartoned material. Specifications for expendable pallets should be available to purchasing from the materials handling department. Purchasing can include copies of these specifications with each purchase order.
- Preshipment testing should be required whenever the buyer feels that it might be necessary.
- The packaging manual should contain corrugated container specifications and include double-wall and triple-wall corrugated container specifications for use with dense material shipments, including fasteners and small parts.

Corrosion prevention   Parts with corrodible surfaces that might rust or otherwise deteriorate should be protected in some way. Some of the things the packaging manual should spell out are, as follows:

- The particular type of cleaning method that should be used for the part should be specified.
- The part should be cleaned before the application of a preservative.
- Some parts might not require either cleaning or preservative treatment; therefore, experience should be the guide.
- The exact method of applying the preservative coating should be indicated, viz.: dip, spray, or brush
- Vapor-controlled inhibitors (VCIs), sometimes called *volatile corrosion inhibitors*), which are chips made of paper impregnated with VCI and VCI granules, both of which release vapor which is capable of coating a part with a film to extend the shelf-life of the part, should be specified. To be most effective the container should be tape-sealed or glued shut. The volatile inhibitor then forms a light film over the part, and even if moisture is present the part is not affected for lengthy periods of storage. Clear plastic films that can be heat-sealed also will provide a measure of moisture protection for most packaging needs, and steel and brass parts can be coated with light oils before the plastic is sealed. Desiccants like silica gel also can be added to a clean part to trap any ambient moisture in the parcel.
- It is also beneficial to include ultraviolet inhibitors for certain materials. Opaque plastic films, usually of a black color, will handle most ultraviolet deterioration. In addition, by specifying the heat-sealing of plastic package liners, in specific instances of sensitive and high-dollar-value parts, it is possible to minimize the effects of oxidation and corrosive fumes. The strength of the container and exterior characteristics of the packaging will depend, of course, upon the product requirements and the expected shelf-life of the item.

Returnable vs. nonreturnable containers   Although there may be two schools of thought on the subject of returnable vs. nonreturnable containers, in my experience there is a place for both types of containers in most companies, depending on the circumstances. Also, it is possible to see many different styles and shapes of each type being used, even in the same company. (For further discussion of containers see chapter 9.)

The packaging manual should include specifications for both types of containers based upon the company's experience. Also, since there are advantages and disadvantages to either type of container, the purchasing department should request the materials handling group to indicate by a cost-benefits study which type of container would be suitable. This should be done only when there are many multiple shipments involved or when the purchasing contract period extends over a considerable period of time.

As a general rule of thumb for motor freight, the point of diminishing returns for most returnable containers is a 300-mile radius of the plant. Beyond this radius, freight rates for over-the-road carriers—i.e., commercial carriers—weigh against the use of returnable containers. On the other hand, "tender" or preferred rates sometimes can be obtained from motor freight companies that will make the use of returnable containers between the supplier and the company attractive.

When considering the use of returnable containers, certain features are desirable, such as:

- A low tare weight
- Capability of being shipped knocked-down (KD), collapsed, or nested

In addition to these characteristics, one of the objectives in the use of steel containers is to obtain maximum truck payloads of at least 40,000 pounds. Anything less would be considered LTL and would bear a considerably higher freight rate.

Railroad returnables   If the company has the good fortune to have a railroad siding, a place should be reserved in the packaging manual for the use of what is known as *railroad returnables*, those containers that the railroads will return to a supplier at no cost to the user. An additional advantage of railroad freight haulage is that the effective range of this logistic stratagem is much greater than the 300-mile radius ascribed to motor freight.

Container accountability

Most manufacturing companies have a considerable investment in containers, and since the movement of materials, by and large, takes place in containers, the responsibility for keeping an inventory of company containers is largely a responsibility of materials management. Delegation of memo charges and accountability becomes a primary concern of the receiving and accounting departments.

If returnable containers are the property of the supplier, his or her invoice will indicate the cost per container. This will be either an actual charge, or memo charge; therefore, if the container is destroyed or is being used in-house, then the customer (your own company) must pay for the cost of the container. Unfortunately, when returnable containers are made of wood, as opposed to steel, sometimes they might not be found in the plant because some unthinking or unconcerned individual has destroyed them, discarded them, or thrown them into the incinerator. Multiply this occurrence by the number of containers unreturned, and the cost might be staggering. Thus, constant vigilance, education, and, primarily, controls are required to stop this senseless waste.

On the obverse side of the coin, however, there are some difficulties that might make them less desirable to use than expendable packaging. For example, they usually require a large amount of storage space and often are of such poor quality that they are hardly worth returning. Containers that are returned KD, or collapsed, require a great deal of expensive labor effort to knock down; and if they can be knocked down, it is possible to do so only with considerable effort, especially after the containers have been in use for a while. They also become *increasingly* more difficult to work with. The older KD containers have a tendency to lose their original dimensions, and the initial configuration that permitted easy assembly no longer remains. In many instances, it becomes virtually impossible to set them up or collapse them, and the tendency is to let them remain in the assembled mode.

Company-owned containers and supplier containers require some form of legible identification; otherwise these containers will be misused. If the company has captive containers that it shuttles between suppliers, it is necessary to have each container clearly marked with the following information:

- Company name
- Return to (location)
- *Returnable Container* in large letters on more than one side of the container.

If the returnable containers are supplied by the vendor, then he or she should give the same information on all packing lists and invoices. Another important note for the purchasing department: it should be clearly explained to each supplier that if not all of the returnable containers are marked properly, then deposits or payments for the containers will not be made.

Where company-owned containers are used in exchange agreements with suppliers, all suppliers in the program should be advised of the types/styles and sizes of containers that will be made available to them. Also, the company would do well to standardize container types so that there is no proliferation of many types and sizes of containers in the company. Also, if the company has a number of plants and the containers are used interplant, a container standardization committee in the materials management department, in particular coordination with the materials handling groups, should affirm that storage space, mobile, transportation equipment, and materials handling equipment are compatible with the unit loads being considered.

Periodically, some one person from the materials management department should visit supplier plants to make certain that company-owned returnable containers have not found their way into the supplier's materials handling system. One supplier's plant, a stamping plant, that I visited, was using several hundred of the company's steel, returnable containers in the plant and storage yards for their own materials. At an average cost of $60 per container, this represented a diversion of approximately $20,000 of the company's capital. When this practice was noted, a strict container accounting system was established to minimize this problem. The entire container accountability program that was instituted, and the additional accounting cost, represented only a small fraction of the cost of containers in use at the company, or about $.03 per container per year.

Identifying receivables

Since the flow and movement of materials are primary concerns of materials management, it is only logical that an effective receiving operation is highly desirable. Thus, in order to move materials through the receiving department in as quick and orderly a manner as possible, the materials management department must do everything practicable to make it easy for personnel to identify incoming shipments quickly and readily. Therefore, the company's packaging manual should contain specific instructions as to what is required on all supplier shipping labels and tags, as follows:

- SHIP TO; (plant name and address)
- PART NO.: (give the complete part number)
- QUANTITY: (number of pieces in each package)
- P.O. NO.: (purchase order number)
- MODEL/CHANGE NO.: (if a model number, give the latest model number; if an engineering change number, give the latest change number)
- DATE PACKED: (show the date by month, day, and year in which the material was packed; this information is necessary for the turnover of inventory and stock control)
- WEIGHT: (gross weight in pounds and metric measure)
- SUPPLIER'S NAME: (the supplier's name, and shipping plant address if a multiplant supplier is involved)

Labeling methods should be standardized as much as possible. There is a threefold impact in this methodology: it speeds up processing on the receiving end—your end; it gains respect on the part of your employees that your company has some clout with suppliers, and therefore, this is good for morale because you are doing it "by the numbers;" it makes the supplier's employees treat your company's materials with a little more respect than they might otherwise do.

Therefore, a form should be included in the packaging manual that would indicate your company's label and tag requirements. Be as specific as possible by showing the actual dimensions and layouts of all identification labels and tags so that the supplier will know without question exactly what is required. Although you are trying to make the means of identification quick and easy for your receiving personnel, you should not penalize the supplier by insisting on a printed label, especially if it should happen to be a small supplier. You should make it perfectly clear that handwritten labels are acceptable, but only under the following conditions:

- The labels and tags can be hand-lettered in a legible manner with waterproof inks (or otherwise protected)
- Tags, when used, should be of 110-pound card stock, or heavier

Sometimes it is necessary to add special instructions to the unit load to ensure the safe arrival of the materials. These instructions should not be a part of the shipping label, but should be marked on the load in the following manner:

- SPECIAL HANDLING—In order to prevent damage, these words should be in large letters at least 2 inches high.
- DO NOT BOTTOM TIER—Tiering instructions also, in letters at least 2 inches high, should be printed on the load to prevent other, and sometimes heavier, loads from being stacked on your fragile materials in transit, especially with LTL shipments since less than truckload quantities are usually consolidated at various trucking terminals.
- DIRECTION OF TRAVEL—This instruction is especially important for certain types of expendable, block-design pallets that have greater strength in one direction than another at right angles to the first. Bottom deck boards of this type of pallet should always run parallel to the direction of travel of the load in transit.

Bills of lading and packing lists

A bill of lading (B/L) is a receipt from a common carrier concern for goods accepted by the carrier for transportation. Although most of the printed material on these bills is composed of boilerplate clauses having to do with carrier regulations, it is advisable that the usual information carried on the bill, such as name and address of consignee and consignor, should be supplemented by including the number of units, or packages on each skid or pallet.

In addition to the B/L, a packing list (P.L.) is a necessary and essential part of each shipment. Without a packing list, checking receivables becomes a very tedious, time-consuming, and utterly irritating experience. The P.L. is important to have with the receivable because it gives the receiving clerk the information needed as to what has been shipped and what is received. The packaging manual, therefore, should spell out what is required in the way of a P.L. as follows:

- Every shipment should have a P.L.
- The P.L. should be placed on the receivable in such a manner that it can be found easily, not buried in the load.
- The P.L. should be visible on the outside of the load, or container.
- The P.L. should be placed on the side, or end of the load, not on top of the load because if another load is tiered above it, the P.L. will be lost or destroyed.
- Whenever possible, the P.L. should be placed on the load that is nearest the door of the carrier (in the main, this would apply to full truck loads).
- Whenever possible, the special red-bordered plastic-windowed P.L. envelopes should be used. They have a pressure-sensitive adhesive on one side. They will protect the P.L., and since they are labeled *Packing List*, they are highly visible.

The packaging manual also should include the information that should be part of every packing list, as follows:

- Supplier's name and plant address
- Purchase order number

- Part or model number
- Engineering change number, if applicable
- The receiving plant's name and address
- The number of units or packages per pallet or skid
- The number of pallets, bulk containers, or skids in the shipment
- The number of pieces in the shipment
- The number of pieces ordered, and/or number of pieces backordered, if any
- A description of the material, if applicable
- If returnable containers are included in the shipment, the information indicated under the section "Container Accountability" earlier in this chapter should be added to the P.L.

According to most state and federal regulations, the supplier is responsible for the initial loading of the materials into the common carrier. From that point on, the supplier and the carrier share joint responsibility for the method of loading materials into the load-carrying vehicle when the materials are unloaded and then reloaded at transfer terminals or freight consolidation locations. Thus, the carrier should report on the B/L all observations and comments concerning aspects of incorrect loading by the supplier when the materials are received at its transfer consolidation terminals.

The transportation element

In small companies, the purchasing department might handle the traffic function, but in a fair-sized company, and in all large companies, the traffic details are handled by a separate group of individuals. Therefore, the packaging manual should indicate that one of the responsibilities of the traffic department is to specify the method, the type of equipment, and the routing for purchased materials. The transportation mode will depend to a large extent on the size of the shipment, the urgency with which the materials are required, and the kind of material to be shipped.

Since friendships are often established between truckers and your traffic people, the materials manager should oversee the fairness of a rotation system, where all eligible and qualified carriers are given their fair share of the company's trucking business. The materials manager should obtain a monthly printout from the traffic department listing carriers by name with destination points and tonnages so that any possibility for collusion is inhibited.

Good dispatching and traffic management can be a real money-maker for most companies, and good traffic personnel combine art and science in obtaining the lowest transportation rates for their companies. Experienced traffic people are able to take advantage of freight classifications that will provide the company with the lowest possible rates. Two of the bibles used by good traffic persons are:

- Motor freight "National Motor Freight Classifications (NMFC)—Classes and Rules"
  H.J. SONNENBERG, Issuing Officer
  1616 P Street, N.W.
  Washington, DC 20036

- Railroads "Uniform Freight Classification (UFC)—Ratings, Rules and Regulations"
  J.D. SHERSON, Tariff Publishing Officer
  Room 1106
  222 South Riverside Plaza
  Chicago, IL 60606

Hazardous materials

An item is considered *hazardous* if it is corrosive, explosive, flammable, radioactive, toxic, or packaged in a container that is considered to be hazardous, such as a pressurized, aerosol-type container. As such, hazardous materials might be difficult and troublesome for many smaller companies, where these materials are infrequently handled or received.

It is the responsibility of every supplier of hazardous materials to comply with all federal, state, and local laws governing the shipment of hazardous materials. The packaging manual should contain the requirement that each supplier supply a hazardous materials safety data sheet prior to, or with, each shipment. If the data sheet is sent to the company prior to the actual shipment, it should be sent directly to the appropriate buyer.

A certain familiarity with government regulations is important and helpful; thus, the materials management department should be sure to conduct training sessions for all plant personnel who are handling hazardous substances. Several helpful sources of information are:

- "Hazardous Materials Regulations of the Department of Transportation" including "Specifications for Shipping Containers"
  Issued by: R.M. GRAZIANO, Agent
  1920 L Street, N.W.
  Washington, DC 20036
- "Motor Carriers Explosives and Dangerous Articles"
  Issued by: AMERICAN TRUCKING ASSOCIATION, INC.
  1616 P Street, N.W.
  Washington, DC 20036
- "Official Air Transport Restricted Articles"
  Issued by: AIRLINE TARIFF PUBLISHERS, INC.
  1825 K Street, N.W.
  Washington, DC 20006
- "Hazardous Materials—Emergency Response Guidebook" (DOT P 5800.2)
  MATERIALS TRANSPORTATION BUREAU
  Research and Special Program Administration
  U.S. Department of Transportation
  Washington, DC 20590

For an overall view of the subject, you might want to refer to my book *Handling and Management of Hazardous Materials and Waste* (New York: Chapman & Hall, 1986).

Packaging deficiency reporting

When contents of materials containers spill out over the receiving floor or leave a trail of spilled materials as the forklift truck backs out of the carrier and into the plant, there are a number of unhappy people who are assigned to clean up the mess, and plant morale suffers. A method for resolving the recurrence of episodes such as this lies with the material management department in the form of a Packaging Deficiency Report (FIG. 10-3), which should be included in the packaging manual. Therefore, whenever materials arrive at the plant in an unsatisfactory condition, or if they have been packaged in a manner that does not conform to the guidelines of the packaging manual—for example, if they are too large, too heavy, or too small—then a deficiency sheet should be completed by the company's materials handling or packaging engineer and sent to the supplier by the appropriate buyer.

Another way of giving the company's packaging program a good deal of support and impact is to require that any deviation from the packaging manual be approved by both the purchasing department and the plant materials handling engineer. It would seem reasonable to permit exceptions to be made on relatively rare occasions and when emergency shipments are required.

PACKAGING & DEFICIENCY NOTICE

Date Rec'd _____   Supplier _____   Date
                                   Name and _____   of Notice _____
                                   Address _____

Part or Model No. _____

Gentlemen: The packaging of a recent shipment received from your company resulted in the arrival of the materials in poor condition. Please complete and return two copies of this form indicating the action you have taken to remedy this situation.

               Buyer _____ Location _____
1. Unit Load:
   _____ Incorrect size for storage L × W × H
          Load size we require L _____ W _____ H _____
   _____ Material loaded in carrier poorly.
   _____ Part numbers mixed. Please consolidate.
   _____ Excessive carton void spaces.
          Pallet too small _____ too large _____
   _____ Container failed
   _____ Improperly marked
   _____ Banding defective
   _____ Other

Fig. 10-3. Packaging deficiency reporting form used to describe unsatisfactory supplier packaging.

2. Identification/Packing List:

_____ Improper location

_____ Not available

_____ Incomplete information

_____ Other

3. Transportation:

_____ Unit load failed because of improper loading/handling

_____ Fragile items not protected

_____ Dunnage between tiers missing

_____ Excessive weight on bottom load

4. Pre-shipment Test:

_____ Container/unit load failed

_____ Packaging failed

_____ Pre-shipment test recommended

_____ Certified pre-shipment test required

5. Comments: _____

_____

_____

6. Supplier Action Taken:

(To be completed by supplier)

_____

_____

_____

_____

Packaging/Material Handling Engineer _____

Photo Attached ☐

Fig. 10-3. Continued.

11

Master schedules and the materials management function

The master schedule

The development of a company's master schedule is the result of a logical sequence of events that, in most companies, begins with the sales forecast. The potential sales figure for each product is derived from a combination of elements that evolve from marketing research studies, actual orders from customers, and the judgment exercised in this area by the senior executives of the company. It is a heady process, to say the least, and many a forecast winds up being fine-tuned by the "experienced" intuition of someone in the executive suite, usually the CEO, saying "let's increase the quantity by 5,000 units."

Fortunately for the materials manager, he or she is rarely concerned directly with the sales forecasting process. However, he or she is definitely a part of the group of people that translates this forecast into a master production schedule. Often the company's scheduled output is exactly the same as the marketing group's forecast, although it might not always coincide with this figure. Nevertheless, the basic production schedule must be as realistic as possible; therefore, it must be responsive to the demands of the marketplace and a number of other factors. The satisfactory schedule should be subject to a minimal number of revisions in order to produce the most economical type of factory operation, such that all of the basic elements of materials, plant capacity, and availability of labor are taken into account.

Since the master schedule has a profound influence over the operation of every department of the company, its final preparation and approval are almost always the result of a committee endeavor. The composition of this committee is a key to materials management effectiveness, since representation in the group indicates the status of this organization. Thus, in addition to the materials manager, there should be representatives from finance, manufacturing, marketing, and business economics. All members should be on the managerial level.

In chapter 6, the methodology of MRPI and II indicated the importance of exploding the bills of materials and deriving the net requirements used in the master schedule. The computer into which these data are entered should, at all times (real-time), reflect the latest information on the demand and inventory status of the parts involved. Similarly, all changes in production schedules should be duly noted, as well as all receipts or withdrawals of materials from storage. Having this up-to-the-minute information helps to smooth out some of the difficulties that arise occasionally in the scheduling area.

Despite the fact that they can be, and are, changed, schedules exhibit a certain amount of inflexibility by the very nature of the products they describe. For example, individual parts, components, and raw materials from which the parts are fabricated must be ordered weeks, and sometimes even months, before it is time to assemble the final product. Always of concern during this period is the possibility that the schedules will be changed, simply because the demand for the end product may vary. It is usually easier to adjust a schedule upwards from the suppliers' standpoint. Scaling quantities back, especially if raw materials are in production, can cause major disruptions and involve cancellation charges.

Not only are these types of changes difficult and time-consuming to handle, but they might upset other schedules for production parts and components. Another difficulty that might cause disruptions up and down the line is the vexatious production problem that might arise because of faulty materials, poor quality control, or a host of other troubles.

The rank-and-file employee who is performing largely clerical functions of a routine nature is quite capable of calculating requirements when everything is going smoothly without hitches; however, when the unusual happens then it becomes the task of the materials manager, or his department, to lock horns with the problem.

In the course of maintaining control over hundreds, perhaps thousands, of items in the production process, there are bound to be certain items that will test the managerial skill of the materials management organization. In addition to quality problems, shortages of raw materials, tool inadequacies, and so forth, there might be supplier problems, such as strikes or work stoppages.

In the materials management view, planning and control have a crucial impact on the optimum levels of operation that can be achieved in a manufacturing system. Among the preproduction activities that define what is to be produced by the factory, the quality of the master schedule significantly affects the capability of controlling what actually takes place on the shop floor. (This is also true, to a large extent, in non-manufacturing enterprises.)

Factory scheduling

The material resource planning techniques that were discussed in chapter 6 are required to be reduced to the practicalities of the shop floor or to the working levels of nonmanufacturing organizations. Thus, we have the factory scheduling process, which is a method of providing the company's manufacturing, fab shops, and assembly lines with the optimum quantity of materials required for efficient operation. Hence, the emphasis on materials management, per se, because less than the optimum quantity of materials will, more than likely, increase production costs and often might result in costly shutdowns. On the other hand, more than the optimum quantity of materials or parts supplied to the factory will increase storage and materials handling costs, in addition to the amount of interest required to be paid on capital invested in inventory, since this money could be more gainfully employed in the reduction of debt, or debt service.

In this fashion, then, factory scheduling, when successfully performed, is of immense value to the company through its ability to save money by eliminating unnecessary machine setups and by safely maintaining inventories at a minimum level. Properly sequencing jobs and operations, together with producing parts in multiple-month quantities, might eliminate a number of extra setups. However, it is important when considering this aspect of manufacturing production to realize that multiple-month production runs depend to a certain degree on engineering changes in the offing, and this factor requires excellent liaison between manufacturing and engineering.

Another balance that must be achieved is the one between manufacturing and inventory control, the question being, when does inventory reduction outweigh the benefits of eliminating machine setups? This is a question of scale that must be resolved between the two departments of manufacturing and materials management. In-house formulas, or guidelines, should be established for the scheduling function wherever it is possible to do so.

Since it is the group that deals directly with almost all of the plant's organizations, the materials management organization should strive, in its coordinating capacity, to serve as a coordinating activity between manufacturing and engineering in implementing engineering changes in existing end products. One of the major emphases, therefore, in holding down the number of setups is to coordinate engineering changes by precisely controlling these changes, through the proper coordination of material procurement and the subsequent machining and fabrication of parts that will produce an end product as required by the schedule.

The factory scheduling group receives scheduling requirement data from inventory control and incorporates this data into an effective machining/fabricating schedule. The information from this schedule is used to compute the requirements for purchase requisitions. The information is then available for the inventory control group to complete the requirements and transmit the data to purchasing. In the purchasing department, the buyer procures materials based upon the quantities and data indicated on the requisitions.

It is the scheduler's responsibility to check status on parts availability periodically to ascertain that the schedule is on target. Through the offices of purchasing and inventory control, the materials manager can obtain periodic reports in order to determine whether any adjustments must be made to meet changing requirements.

Preparing scheduling data

The scheduled operation is usually the one performed by the machine or group of machines with the heaviest burden. *Machine burden* is expressed in machine hours of work required each month based upon a work order setup and machining times. Since the understanding of what constitutes machine burden is relevant to this subject, a brief explanation follows.

Accurate burdens must be computed from a scheduling chart or a manually derived status card, which is kept up to date by being reviewed periodically. When an up-to-date chart is called up on the computer screen, it should cover all requirements available. The burden hours are computed in the following manner: If the production line is scheduled on a five-day workweek, then the percent of capacity is multiplied by 504.

$$21 \text{ days/mo.} \times 24 \text{ hrs/day} = 504 \text{ hrs/mo.}$$

Giving due consideration to man/machine efficiency, three-shift operation, five days/week is understood to be fully burdened at 400 hrs/mo.

If the production line is scheduled on a six-day workweek, the percent of capacity is multiplied by 608, derived as follows:

$$21 \text{ days/mo.} \times 24 \text{ hrs/day} = 504 \text{ hrs/mo.}$$
$$4^{1}/_3 \text{ Saturdays/mo.} \times 24 \text{ hrs/day} = \underline{104 \text{ hrs/mo.}}$$
$$\text{Hours/mo. in a six-day week operation} = 608$$

If the burden hours have not already been stored in the manufacturing database, then they must be manually derived from status cards, using the following methodology. A time period that covers one cycle of machine operation is selected. As an example, if material is run in two-month lot sizes, then a two-month period should be used. Requirements for a part, over the selected time period, are divided by the pieces per hour (from the work order time) per machine, multiplied by the number of machines that are available. The result is the number of machining hours required for the part on each machine.

The three parts listed in TABLE 11-1 are routed over two automatic lathes for the first operation. Each lathe performs the complete operation, and one-month lot sizes are run. The problem to be resolved is the computation of the average monthly burden per machine for a period of three months. On the worksheet, which might consist of a computer program, set up the tabulation in TABLE 11-1.

As you can see from this tabulation, the anticipated number of setups for each machine producing the part is multiplied by the work order time for the setup, and added to the machining hours required to produce the part on that machine. The sum of the burden hours for each machine for all parts routed over the machines to be used is then divided by the number of months in the time period being considered to obtain the average burden per machine per month. It is important to note that when burdens over more than one machine are computed and all parts are not routed over all machines, the burdens over the several machines should be balanced as closely as possible so that shop floor operations also can be balanced effectively.

In a well-run and well-organized plant, whether it be of a manufacturing or non-manufacturing enterprise, work pace usually can be readily determined by the astute observer. Although it is somewhat more difficult to make this determination in non-manufacturing plants, there is always some method available for gauging this effective-

Table 11-1. Burden Hours per Part.

Part No.	Pieces Req'd	Pcs/ Hour/ Mach.	No. of Machs.	Mach. Hours	Set-ups/ Mach.	Hours per Setup	Setup Hrs/ Mach.	Burden Hrs/ Mach.
8Q 9120	4,000	20.5	2	4.000 20/5 × 2 = 97.6	3	7	21	118.6
6P 2841	14,000	8	2	14,000 8 × 2 = 875	3	5	15	890.0
4S 6395	6,500	18.5	2	6,500 18.5 × 2 = 175.7	3	6	18	193.7

Burden hours total = 1202.3

Average monthly burden hours/machine = 400..8

ness, and it rests solely with the manager of the operation and his or her staff to determine what these measures are.

In the manufacturing plant, a condition might arise that could very well be a concern of the materials management organization and one that might cause some difficulty in resolving, especially in times when budgets for each department are severely straitened and hence cannot stand the added strain of unexpected and unforseen events. As an example, let us assume that the manufacturing department has been performing well, and is on target as far as the master schedule is concerned. Also assume that the work pace is at an acceptable level of performance and that scheduled machine maintenance is keeping the machine running at an up-time rate that is well within the expectations and standards established for each group of machines. As it happens, one of the factory schedulers becomes aware that he or she has an overburdened machine— i.e., one that is unable to produce current requirements by working three shifts daily, five days per week, or approximately 400 burden-hours monthly.

Since this condition can be disastrous in terms of the disruptions it might cause if left unresolved, a very structured and formal procedure is necessary to impress the shop foreman, the scheduler's supervisor, and, of course, the scheduler on the importance of taking immediate action. Thus, the scheduler, on observing that one of his or her machines is overburdened, must alert the shop foreman immediately in order to determine the course of action to be taken. This step is necessary, especially in some plants where work is not based on a three-shift operation. The scheduler in this type of plant must make allowances for this fact and scale down the 400-hour burden rate to a more acceptable level. The scheduler can then do his or her part by helping to balance shop-floor operations more effectively. If, on the other hand, the scheduler's plant is on a three-shift, five-day workweek, then he or she must formally pursue the overburdened condition by documenting the extent of the overburden, giving its cause, and offering a recommendation for its solution.

A standard operating procedure should be developed for overburden "alerts," which would consist of the following measures:

Scheduler:

- On a preprinted form explain extent of overburden.
- Explain cause.
- Make recommendation for overcoming overburden.

Supervisor:

- Take action to overcome overburden.
- Review scheduler's measures.
- Prepare written report indicating what measures have been taken.

If a burden of 480 hours or more per month occurs (which is the equivalent of a six-day workweek), then measures such as the procedure just noted should go into effect. To rectify the overburden, one or more of the following steps, which are of particular significance to materials management, should be taken:

- Authorize Sunday work.
- Purchase machine time from a subcontractor.
- Purchase additional machine tools.

Occasionally, a machine line that is unable to produce all of the desired requirements might be scheduled below requirements, but slightly above the quantities that it can actually produce in order to keep the schedules from becoming completely chaotic. Again, S.O.P requires that the scheduler's supervisor and higher plant officials be kept informed of the necessity for doing so. It is an interim procedure at best, which offers only a short-term solution until such time as a better action plan can be placed into effect.

The machining cycle

The *machining cycle* represents the most practicable order in which parts should be routed over a production line. A good line foreman should be able to tell the scheduler the best sequencing of machines to be followed. Also, an experienced scheduler might have to adjust this sequence somewhat based upon setup time and the on-hand inventory for the part(s). *Lot sizes*, or the quantity of each part machined in the run, are also determinations made by the scheduler and the scheduling section. The scheduler, working from a master schedule, is required to compile data—which he or she can readily do by storing/retrieving the data bank so that a very precise decision can be made on lot sizes—since there are practical as well as economic considerations that impinge upon the generation of lot sizes.

Processing time

Machining lead time, or processing time, is the time required for the first good piece of each production run to proceed from the scheduled operation through all subsequent operations and inspection checks as a finished piece part. Lead time for the part, most generally, should be based upon a five-day workweek.

The basic production line lead time is applicable to the parts that are to be assembled into the final end product. Therefore, the lead time required for assembly has to be added to this machine time. In addition to his or her scheduling activities, a good scheduler is required to know what work orders have been completed, have been started, and are in process, or have not been started, on his or her production line(s) at every minute. Thus, it helps to have a computer terminal available to the scheduler so that he or she can keep posted on what is happening on the shop floor. The shop floor control system should be initiated by the materials management department so that the movement of materials on the shop floor can be tracked readily for the scheduler's benefit, as well as for the benefit of the shop supervisors.

An added advantage of a good, computerized shop-floor control system is that it permits a good deal of work measurement data to be collected innocuously. Since the information required by the scheduler is readily available in a computerized shop-floor control system, there is very little effort required on his or her part to obtain timely status reports on shop floor progress so that at least daily checks on the output of each machine in the scheduler's line can be made.

A part of the preparatory work that each scheduler must do is to ascertain how much material will be available for the first scheduled run of each part number under his or her control and how much material the purchasing department has on its due-in listing. If there are any changed requirements, or if machining dates have been revised, these factors must be discussed with the materials management department, of which purchasing is the cognizant entity.

Rapport between the scheduling organization and materials management should be particularly close because it is up to the supplier to ship all of the material required for the lot size selected to the plant sufficiently in advance of the machining data so that the material can be properly inspected for use by the company. Depending upon how well the materials management department has organized the receiving function, the scheduler's materials should arrive anywhere from two to ten days prior to the actual machining date.

In coordination with the scheduling section, the materials management department should strive to narrow the time interval between the date the machining is to start and the date the material is received by the plant. The vaunted Just-in-Time Manufacturing has laid claim to narrowing the gap between the receipt of material and its use to a matter of a few hours. This lag in the time interval should certainly not be more than 24 hours, according to some staunch advocates and proponents of the system. However much the gap may be, the implication for materials management is quite clear: the smaller the time interval becomes, the less inventory is required, with an appreciable amount of savings that can be generated by this one materials movement principle.

12

Production control

Introduction

In chapter 1, the organizational structure of a manufacturing enterprise was discussed, and an attempt was made to show some of the interrelationship that exists between several of the departments, including materials management, that comprise the modern factory. Although the emphasis on materials management has been largely one of manufacturing in this text, there are many parallels that exist between manufacturing and service, or nonmanufacturing businesses.

The principles that govern the movement and flow of materials, for the most part, are homologous. Therefore, by concentrating on the manufacturing plant a clearer picture will emerge concerning the materials management concept that will carry over into other commercial, industrial, and nonmanufacturing segments of the economy. The partial proof of this reasoning is that even nonmanufacturing environments such as insurance, banking, and real estate, refer to new services as "products."

The new product

Thus, a new product can be created by design or developed by market research, depending on where or how the concept originated. Nevertheless, despite the origin of the concept, it is up to the engineering department to put together a product description that will delineate the general specifications of the product sufficiently to permit tooling costs to be estimated by the manufacturing division. In nonmanufacturing establishments, the same parallel exists, in that the home office or headquarters staff would be empowered to develop the final specifications for the new "product." Suffice to say that, in this chapter, this will be the last reference to this sort of parallelism.

Determining costs

Before realistic tooling cost estimates can be made, the question of scale must be resolved because the quantity of the product to be manufactured might influence tooling costs enormously. As an example, if a prototype is to be made, hand-cobbled tooling might be entirely satisfactory, whereas if large production runs will be required, then expensive tooling costs might very well prove to give the best per-unit costs over the life of the program.

In some companies, especially multinational enterprises, the internal organizations are so large and structured that they might enjoy the luxury of a business economics department. Mid-sized companies usually have sales and marketing departments. Thus the impact of determining market share for the new product would lie in the purview of these departments.

Somewhere in the maws of these organizations will be the expertise to determine the price of the new product; in other words, how much can the company charge for the new product and still sell it and capture its potential share of the market? Sufficient legwork is performed by staffers in engineering, accounting, and purchasing (as representing materials management) that the approximate cost to produce the product can be known. Enough information is now available, including the R.O.I. for the product, that the product committee, usually composed of the heads of all company departments and their principal staffers, is ready to determine the advisability of adding the item to the company's product line. After a presentation of the proposed product from the marketing department, the product committee makes its recommendation to the company's executive board, or in some instances, the president or CEO, who has the final say in the matter.

Design of the product

If the approval to go ahead with the project is given, then the engineering department will give its best estimate for the date that all the drawings for the new product will be completed. The manufacturing department will then firm up a date that includes production tooling lead-time.

After the engineering design work has been completed, drawing notices and blueprints of each piece part for the product are sent to the planning records group. The personnel in this group establish and maintain control of all work in the planning section; therefore, they can make up-to-date status reports on progress of the new product, whenever required.

The planning section

There are four major areas, or stations, in the planning section: releasing, assembly layout, processing, and industrial engineering time study. Their functions as they impinge on the materials management effort will be defined in this manner because what they do in these several groups affects the flow of materials through the manufacturing process. Therefore, in view of the systems approach, whatever they do will in some way, directly or indirectly, affect the ultimate objectives of the company, which is to make a quality product, at the least total cost, with the maximum profitability.

Station one: Releasing

- Determine the procurement status of each piece part in terms of:
    - Whether it is a worked part (W); i.e., fabricated in the shop.

○ Whether it is a purchased finished part (PF); i.e., purchased in a finished condition with no added fabricated time required. Ball bearings are an example.

○ Establish the source of supply (SOS) and the effective date of delivery.

• Establish and maintain control of sizes, shapes, and material requirements for all piece parts composed and made from raw/unformed materials.

• Order and control patterns, jigs, fixtures, and related tooling and equipment used in the purchase of rough and finished castings, forgings, and purchased finished parts.

• Prepare and maintain a material and station list record of manufacturing data and specifications on all piece parts, assemblies, groups of components, and arrangements.

The effective dates for a product and its spare parts and accessories are transmitted daily to the business economics department through a facsimile machine. This data, together with order rates, inventory position, productive capacity, and the like, are used to prepare a monthly (or biweekly) executive production schedule.

When this data is transmitted to the business economics group, if there are no further questions or concerns, it becomes the authorization to purchase materials and establishes the product fabrication/manufacturing schedules for a period of approximately one year. At this juncture, all purchased finished (PF) items are removed from the overall piece parts package and transmitted to the purchasing department (materials management) in order for this group to establish a supplier base. The items to be fabricated in-house—the worked (W) parts—are then transmitted to the third station, which is processing.

Station two: Assembly layout

• Determine how and where the assembly of the new product will be made:
  ○ Provide the tools and tooling
  ○ Provide space and storage areas along the assembly line
• Issue the assembly work orders, which list, in a sequential manner, all the steps necessary to assemble the product.

In the assembly layout section is a small group called the "assembly" industrial engineering timestudy group, whose duty it is to separate all of the elements involved in the assembly operation into little packages and assign an appropriate time for assembly to each element. This data allows the assembly lines to be staffed to conform to changing product mixes and enables the performance of the line to be computed.

Station three: Processing

• Worked items (W) are received in this station from station two.
• Processors originate *production work orders*, which detail the sequence of operations to be performed on each piece part and indicate the machines and operations to be accomplished.
• Processors issue tool orders for durable tools, jigs, and fixtures required to perform the work order operations.

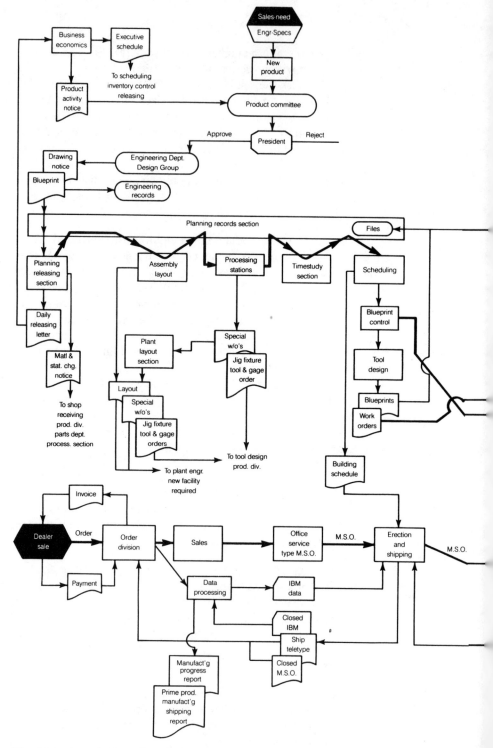

Fig. 12-1. A work-flow diagram illustrating the way in which a new product proceeds

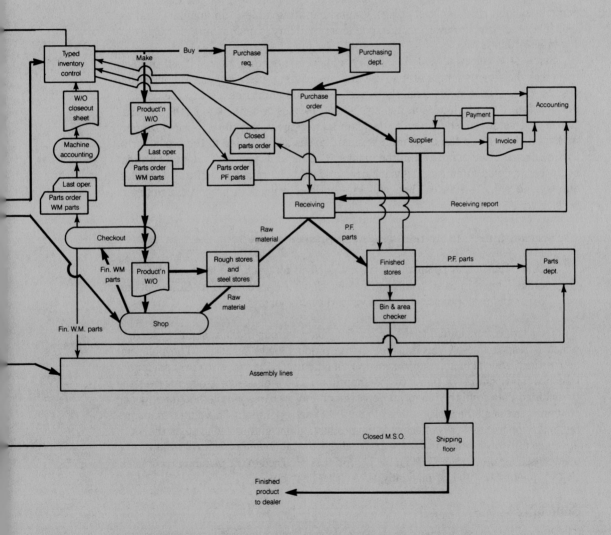

from the conceptual stage to a finished end product.

The tool design section designs or specifies the type of tooling required to produce the piece part. Sometimes the tooling is purchased from specialty houses; often, tooling already built for formerly manufactured products must be—or, more importantly, can be—modified by the tool department's own shop personnel for this new product.

The plant layout section also works with the processing group in order to provide adequate space for manufacturing operations and assists in laying out the workstation arrangement; that is to say, the area surrounding the machine tool(s), which is serviced by an operator (machinist or fabricator-technician).

The layout group, in conjunction with materials management also might originate tool orders. These orders are usually for handling equipment to be used around the workstation and might consist of jigs, cranes, elevating tables, tub dumpers, and the like. The processor, however, has the added responsibility of maintaining adequate tooling in his or her area.

Since new releases, added capacity requirements, and worn-out or obsolete machines that are no longer able to maintain machining tolerances are the inevitable results of production operations, the processor should always be alert to the need for new and improved machine tools and equipment. It is his or her task to justify the addition or replacement of equipment to the plant management; therefore, it would simplify matters and speed up the program if standard forms are provided to the processor so that he or she could simply fill in the blanks. The request, justification, and formulas could be modeled after MAPI [Machinery and Allied Products Institute] formulas, although in some instances they will give inflated values to proposed equipment replacements.

Station four: Industrial engineering timestudy

• Performance and setup times are established for each operation on the work order. These times are used as a base for all piece-part costing, machine tool burdens, and performance comparisons and evaluations.

This data is transmitted to the production control group in the scheduling section, which is responsible for the issuance of the product build schedule. The blueprint department, which maintains a control over all blueprints, then receives the data for the issuance (and control) of prints to the shop floor. The tool design group receives the transmitted data from the blueprint department and prepares job method data, which contains a list of all the tools required to perform each operation. This data is transmitted to the specific tool crib adjacent to the production line authorized to stock the tools. A production work order master is transmitted to the inventory control group to be used in issuing work orders. Then, the blueprint is returned to the planning record section via pneumatic tube (or manually, as the case may be).

Schedules

The business economics group, or its equivalent functional activity, now releases the *executive production schedule* to the inventory control group, which authorizes them to take the following actions:

- Issue purchase requisitions to the purchasing department to procure purchased finished (PF) parts and the raw material for worked (W) parts.
- Issue work orders to begin work on parts that are fabricated in-house.

Work order issuance must be coordinated within the production control and materials management areas of activity in order for the raw materials movement to be accomplished effectively within and through the production shops of the plant. In addition, completed work (W) and purchased finished parts must be available in sufficient quantity to meet the build schedules that have been issued by the scheduling section.

Note that when customer orders are received by the company's order department, the build schedule might be affected if the quantity of incoming orders is higher, or lower, than anticipated. Thus, the level of customer orders that are received affects the executive schedule, which in turn, influences the company's production pattern. If the marketing and sales departments have done their work well, and they have been lucky, then there will not be any severe reverberations throughout the manufacturing system.

Of course, there are never any such things as certainties in the real world, and market conditions can sometimes perform in a manner that defies even the most professional marketing experts. Nevertheless, the modern manufacturing plant, operating as a well-functioning team with a systems-approach philosophy, can adapt itself rapidly to the changing conditions of the marketplace. The tools for this rapid adjustment should be built into every manufacturing and materials management system. Just as the purchasing department might increase order quantities in good times, the cancellation clauses incorporated into contracts, which were discussed in chapter 6, can also scale quantities to manageable proportions when there is a market downturn.

Figure 12-1 is an illustration of the work flow methodology for taking a new product from the concept stage straight through to the production of a finished, end product. The end product in this example is a fairly complex machine, thus the term *machine shipping order* (M.S.O.) is used.

Material control

Since the main business of materials management is to ensure the orderly and effective flow of materials from the supplier, through the plant, through warehousing and storage, to the ultimate consumer at the least total cost, it is easy to see the importance of the material control function in this system. Since controlling material also means keeping precise records, it is possible to distill the functional areas of material control into two main areas: moving material and keeping records.

Keeping good records makes it possible to track materials; therefore, with today's many automatic identification methodologies, such as bar coding and RF tagging, there is absolutely no excuse for materials to get lost in the manufacturing process. There is also no reason for materials to disappear in storage or over the shipping and receiving platforms unless there is collusion between the company's personnel and the trucking company. The use of bar coding as an example, makes the possibility of collusion fairly remote, using the best methods currently available.

Another important element in maintaining adequate records is the question of cost. Since its inception, the price of auto ID systems has trended downward, so no matter how small or large the company may be, auto ID will have an R.O.I. that makes it possible, in most instances, to have a payback in less than a year.

There are a number of reasons why the R.O.I. for bar coding is so attractive. For one thing, you can scale an auto ID system of this type up and down, starting at only a few thousand dollars to many thousands, depending on the sophistication of the system; however, to get up to speed in this technology does not require much of an investment. Any company using a PC computer can obtain a packaged program and the hardware to get into the business. Usually only slight modifications are required to adapt the packaged software to your requirements.

Other advantages of bar coding are that it will enable your company to track incoming materials through the manufacturing process with a minimum of effort. Materials in storage and in production can be counted readily and precisely using scanning wands. If the operation is mechanized, bar code beam scanners can count and track, at very rapid speeds, materials that are conveyorized. Stock clerks can take inventory in either batch or real-time modes, and the counting done by auto ID methods is so vastly superior to manual counting that the degree of precision can hardly be surpassed.

Bar coding, properly used, is very similar to the office copying machine. Companies that have installed copiers and auto ID systems wonder what they would do without them. But there the similarity ends. Bar coding is emphasized here because the records maintained by inventory control and the accounting department depend, in large measure, on the precision and timeliness of the data furnished to them by material control.

The physical areas of the plant that are of concern to material control start with the receipt of material and follow it through the manufacturing process until it is shipped out the factory door. Since the material control group handles large tonnages and vast quantities of materials in terms of piece count, even in the smallest operations the labor and equipment input required by this group makes it an extremely expensive operation. Early in this text the cost of handling materials was said to absorb a large part of the company's dollars; therefore, it is necessary that the plant management be concerned with providing the proper floor layout and equipment to perform the necessary work of moving materials expeditiously.

Materials handling

In the production control group, the majority of employees working in material control are forklift truck operators. If the company manufactures heavy equipment, there might be a number of crane operators, straddle crane operators, and the like. Thus, we are concerned with not only the accuracy of counting materials, but also with the care and operation of equipment and the physical areas of the plant where this work is performed. For the most part, material control personnel are concerned with materials handling, and this group of employees is divided among the functions of receiving, shipping, storage, and in-plant handling.

Although the basic principles of materials handling are common to all of these areas, one of the important tenets of handling is that the operators working within the system be trained completely in the fundamentals of the process. Unfortunately, most training that occurs in many plants is the sort of vestibule, or on-the-job, training that results in careless, slipshod work, simply because the employee feels that he or she is just a number in the macrocosm of the plant and becomes lost in the shuffle. The more formal and structured the employee's training is, the more care and effort will be expended by the individual employee in doing his or her job well. This is a small psychological point that only the larger companies seem to recognize and to which they pay more than lip service. Both their merchandise and equipment are better off as a result of this attention.

The formal training of the materials handler should not stop with the care of the equipment he or she is using, but should be carried on through to the procedures that he or she is required to use during the course of the work shift. Employee morale is at best a tenuous thing; therefore, it must be nurtured, by company management, at every opportunity. What better way to engender company loyalty than to help an employee thoroughly understand the reasons behind the procedures? It is not sufficient to school the employee in the procedures and then let him or her wonder why the company wants such-and-such done in this (peculiar) way.

This is by no means a treatise on employee training, but it is merely laying the groundwork for a sampling of hypothetical receiving procedures that can be applied to almost any company or organization, whether in the manufacturing or service industry.

Receiving purchased materials   If the company is large enough to have a traffic department, then it will be from this group that the purchasing department will obtain traffic information, such as rates, transportation modes, and carrier names. In this mix of transportation methods there will be, no doubt, commercial carriers, both truck and rail; the supplier's fleet; U.S. parcel post; United Parcel Service; ship and barge; as well as city delivery vehicles. Sometimes, the type of carrier or transportation mode will dictate the kinds of materials-handling equipment that might be needed to unload, or receive, the material. Often, with small delivery vehicles or with carriers with poor decks, manual unloading might be the only way to get at the material.

There are three important areas to consider in obtaining the maximum effectiveness in the materials-handling function:

- Physical layout
- Good equipment
- Training

Physical layout   In the first instance, the layout, or arrangement, of the receiving docks and holding areas must be designed so that the maximum efficiency can be obtained from the operation. This is true of any of the areas in which the material-control personnel have the responsibility for the movement of materials. Crowded and cluttered aisles, a lack of turning space for forklift operation, and a great deal of passenger traffic in the working aisles can hamper the operation enormously. Also, inadequate set-out space might result in a forklift operator wasting precious time by having to set a load down in order to stack containers to make room for the load being transported. As you can see, the proper physical layout will enhance the speed and safety of any materials-handling activity, regardless of the area of the facility in which it is being carried out.

Good equipment Concomitant with adequate, well-planned facilities is the need to provide the proper equipment for the material control operator. Included in this statement is the implication that not only is the equipment right for the job it is called upon to perform, but that it is well maintained. Nothing is more discouraging to an employee than to have to use a forklift or other piece of mobile equipment that is leaking oil, that has a slipping clutch, and so forth. This is extremely hard on morale, and it engenders sloppiness and carelessness on the part of most operators.

Training Enough has already been said about training, in a general way. Since this part of the text concerns the procedure originated by the materials management department for receiving purchased materials, it points the way for training material control personnel in a better understanding of their functional responsibilities.

Material control personnel in the receiving group should process receivables in the following manner:

1. Truckers should be instructed to use a particular truck spot on the receiving dock. In well-organized companies, an entrance guard shack should have an intercom station connected to the receiving office. In lieu of an intercom system, an ordinary phone connection would serve as well. Naturally, an intercom outlet should be supplemented with the company telephone system. The receiving clerk dispatcher should assign (a) the trucker to a specific unloading dock, and (b) a material control forklift operator for unloading.
2. The material control person should request the freight bill and shipping manifest from the trucker. (See FIGS. 12-2 and 12-3).
3. The material controller should eyeball the incoming freight to determine whether or not it should be palletized. If it is not, then he or she should obtain a stack of empty pallets and stack them adjacent to the tailgate of the truck. If the material is relatively small in size and should be placed in containers, the material controller should obtain suitable containers that can be moved directly to the production line or placed into storage on the next movement of materials. The materials should be placed into containers that will permit the subsequent operations of handling and storage to be made with the greatest efficiency and still provide protection for the piece parts in transit.

   As part of the materials management philosophy, the unitization of incoming freight is of prime importance in making the receiving operation function effectively and at the least total cost. The purchasing department, in coordination with the materials-handling department, should always specify how the company requires its receivables to be packaged.

   The effectiveness of the receiving function can be enhanced considerably when the vendor is aware of his or her responsibilities in this area. Also, whenever it is possible to institute a container program with a vendor, it is possible to make large savings by eliminating packaging materials that must be discarded after one use. Not only are there ecological benefits by eliminating scrap and

STRAIGHT BILL OF LADING – SHORT FORM – ORIGINAL – NOT NEGOTIABLE

RECEIVED, subject to the classifications and tariffs in effect on the date of ' this Original Bill of Lading.

FROM AT

F. H. LANGSENKAMP CO., INC.
229 East South Street
Indianapolis, Indiana 46225

NAME OF CARRIER **TRANSCON**

DATE 10-5 1987

SHIPPER'S NO. H9113

CARRIER'S NO.

the property described below, in apparent good order, except as noted (contents and condition of contents of packages unknown), marked, consigned, and destined as indicated below, which said carrier (the word carrier being understood throughout this contract as meaning any person or corporation in possession of the property under the contract) agrees to carry to its usual place of delivery at said destination, if on its route, otherwise to deliver to another carrier on the route to said destination. It is mutually agreed, as to each carrier of all or any of said property over all or any portion of said route to destination, and as to each party at any time interested in all or any of said property, that every service to be performed hereunder shall be subject to all the terms and conditions of the Uniform Domestic Straight Bill of Lading set forth (1) in Official, Southern, Western and Illinois Freight Classifications in effect on the date hereof, if this is a rail or rail-water shipment, or (2) in the applicable motor carrier classification or tariff if this is a motor carrier shipment.

Shipper hereby certifies that he is familiar with all the terms and conditions of the said bill of lading, including those on the back hereof, set forth in the classification or tariff which governs the transportation of this shipment, and the said terms and conditions are hereby agreed to by the shipper and accepted for himself and his assigns.

(MAIL OR STREET ADDRESS OF CONSIGNEE—FOR PURPOSES OF NOTIFICATION ONLY.)

CONSIGNED TO AND DESTINATION

Aviation West
Sea-Tac Cargo Center
2580 South 156th Street
Seattle, WA 98158

ROUTE

Delivering Address # *☞ TO BE FILLED IN ONLY WHEN SHIPPER DESIRES AND GOVERNING TARIFFS PROVIDE FOR DELIVERY THEREAT

DELIVERING CARRIER | CAR OR VEHICLE INITIALS & NO.

No. of Shipping Units	Hazardous Materials	KIND OF PACKAGES, DESCRIPTION OF ARTICLES, SPECIAL MARKS AND EXCEPTIONS	*WEIGHT (Subject to Corr.)	CLASS OR RATE	CHECK COLUMN
		BUNDLE ALUMINUM DOCKBOARD NMFC ITEM 168720 - Sub. 1			
1		(1) 6084 DMTS/4	348#		
		ATTN: Mark Develbiss			
		PO# 052392			
		TOOL PLATFORMS — FORK LIFT TRUCKS NMFC ITEM 150390 - Sub. 1			
		DOR-SEALS NMFC #157320 - SUB. 1			

Subject to section 7 of conditions of applicable bill of lading, if this shipment is to be delivered to the consignee without recourse on the consignor, the consignor shall sign the following statement:
The carrier shall not make delivery on this shipment without payment of freight and all other lawful charges.

Per _____
(Signature of Consignor)

If charges are to be prepaid, write or stamp here, "To be Prepaid."

PREPAID

Received $ _____
to apply in prepayment of the charges on the property described hereon.

Agent or Cashier

SHIPPERS CERTIFICATION This is to certify that the above-named materials are properly classified, described, packaged, marked and labeled, and are in proper condition for transportation according to the applicable regulations of the Department of Transportation.
SIGNATURE: | TITLE:

Per

REMIT C.O.D. TO: (ADDRESS)

C.O.D. AMOUNT $

C.O.D. CHARGE TO BE PAID BY { SHIPPER ☐ CONSIGNEE ☐

(The signature here acknowledges only the amount prepaid.)

Charges Advanced,
$

† This is to certify that the above-named articles are properly classified, described, packaged, marked, and labeled, and are in proper condition for transportation according to the applicable regulations of the Department of Transportation.
* If the shipment moves between two ports by a carrier by water, the law requires that the bill of lading shall state whether it is "carrier's or shipper's weight".
‡Shipper's imprints in lieu of stamp; not a part of Bill of Lading approved by the Department of Transportation.
NOTE—Where the rate is dependent on value, shippers are required to state specifically in writing the agreed or declared value of the property.
† The fibre containers used for this shipment conform to the specifications set forth in the box maker's certificate thereon, and all other requirements of Rule 41 of the Uniform Freight Classification and Rule 5 of the National Motor Freight Classification.

THIS SHIPMENT IS CORRECTLY DESCRIBED.
CORRECT WEIGHT IS _____ LBS.

The agreed or declared value of the property is hereby specifically stated by the shipper to be not exceeding _____ per _____

Shipper _Ray Davis_
Per _____

Agent _Transcon_
Per _____ 10-5

F. H. LANGSENKAMP CO., INC.
229 EAST SOUTH STREET
P.O. BOX 1106
INDIANAPOLIS

Permanent post office address of shipper

Fig. 12-2. Freight bill showing number of pieces, shipping weight, and charges.

PICK UP MANIFEST

DRIVER # ____ DRIVER NAME ____ TRAILER # ____ TRIP # ____

SHIPMENTS ____ AREA ____ TRLR P/U ____ IN YARD LOCATION ____ P/U DATE ____ TERM CODE ____

NOSE LOAD ____ DEST TERM ____ NOSE LOAD ____ % CUBE ____ LOADER # ____ LOADER NAME ____ WORK DATE ____ PRINT PRIORITY ____ SHIFT ____

DRIVER'S COMMENTS:

	PRO NUMBER	ARRIVE / DEPART	SHIPPER	PCS	WT	PLT CT	HM	PIC	SPEC. HAND.	CONSIGNEE NAME	CITY	ST	ZIP	CUBE INFO LEN	CUBE INFO WID	CUBE INFO HT
1																
2																
3																
4																
5																
6																
7																
8																

VIKN 0-0435 Rev. 2/90

Fig. 12-3. An illustration of a shipping manifest.

waste, but the method of containerization might improve the handling and storing of the shipped materials. Thus, it is possible to eliminate two extra handlings of materials, and the added effort required to remove the packaging materials from the part and the packaging materials from the site.

4. In a manual receiving system where the only mechanization is forklift trucks and conveyors, the material controller must obtain a traffic number, or the like, from the receiving office before he or she can start to unload the carrier. The traffic number is applied to (all) freight bills, packing lists, and receiving reports, as well as over, short, or damage reports. The traffic number is the reference basis used in tracing shipments and paying for the merchandise.

   The accounting department's accounts payable section should require complete documentation prior to paying a bill; therefore, all documents for a particular shipment must bear the traffic number. This is the principal method for ensuring that the materials paid for have been received.

5. As the materials are being removed from the truck, the material controller will check the items received against the freight bill in terms of quantity of pieces, boxes, crates, cartons, and the like. He or she also should visibly inspect the outer containers to determine if there is any evidence of damage. For example, a crushed carton, broken slats of a box, a torn burlap wrapper, or the equivalent signs of rough handling should be noted on the freight bill.

   The material is then moved into an adjacent receiving setout area or checking bay. After the material is completely unloaded and the container piece count has been verified, the freight bill is signed by the material controller and a copy is given to the trucker so that he or she can drive away the vehicle. As part of the entrance and exit procedure, the trucker should be made to stop at the guardhouse on his or her way out of the factory complex so that the guard can stop the tractor-trailer or truck and have the trucker reopen the tail-gate, if he or she has already closed it. The guard visually inspects the contents to guard against pilferage.

6. If the material controller has discovered any overages, shortages, or damaged materials, it was indicated that these instances were recorded on the freight bill. In addition to documentation on the freight bill, the material controller is responsible for completing the Over, Short, or Damaged Report form. (See FIG. 12-4).

   The material controller is responsible for obtaining the trucker's signature on the amended freight bill; otherwise the proof of the company's claim might be difficult to establish, especially with new vendors.

7. If the material controller happens to be unloading very fragile, complex, or precision-type instruments or machines, the requisitioning department should be notified immediately as the material arrives at the dock so that a representative of the department can be on hand to make a visual or actual examination of the item(s), prior to releasing the carrier.

8. Upon satisfactorily clearing the freight bill and shipping manifest, the material controller turns them into the receiving office, where they are duly recorded and transmitted directly to the accounting department.

OVER, SHORT, OR DAMAGED REPORT						(Form 2159)

Date   File No.

Shipper

Address
Delivering
Carrier   Traffic
                                                 Reference No.

                                                                         PRO
Weight _____ Car No. _____ No. _____

Purchase Order No.	Symbol or Container	Items per Freight Bill	Itens over Count	OVER	SHORT	Weight

F.o.B.   Invoice   Invoice
Point   Reg. No.   Due Date

Damage Report

Traffic Claim No. _____ Final Disposition _____

Fig. 12-4. Example of an Over, Short, or Damaged Report form.

9. The material controller is also responsible for preparing a receiving report. In order to do this properly, the material controller must count the material received and compare this quantity with the supplier's packing list. (See FIG. 12-5.)

Interlake CONVEYORS

P. O. Box 276 • Shephердsville, Kentucky 40165

502-543-7046

PACKING LIST

PAGE	1 OF 1
A.J. BAYER JOB NUMBER	91622 01
JOB TRUCK NUMBER	—
BILL OF LADING NUMBER	15341
DATE SHIPPED	6-21-89
CUSTOMER ORDER NUMBER	VETB/TED ALLEGRI

SHIP TO: D & A ENGINEERING

STREET ADDRESS: 14008 92ND PLACE N.E.

CITY: BOTHELL STATE: WA. ZIP: 98011

MARK NUMBER	QUANTITY	DESCRIPTION	SKID #	BOX #	BUNDLE #
A-1	10	4" E SIDERAIL 6" O/C x 120"			
		1-SKID @ 680 LB			
		ATTN: TED ALLEGRI			

PARTIAL SHIPMENT ☐ COMPLETE SHIPMENT ☑ THIS IS A PARTIAL SHIPMENT THAT COMPLETES ORDER ☐

CARRIER: VIKING F.T.

INSPECTED BY ______ DATE ______

RECEIVED BY ______

SHIPPING DEPARTMENT APPROVAL BY: K. Burgin DATE: 6-21-89

DATE RECEIVED ______

PLEASE CHECK ALL GOODS WHEN RECEIVED (SEE REVERSE SIDE)

DISTRIBUTION: Original — White Customer — Canary Production Control — Pink Shipping — Gold Sales — Green

Fig. 12-5. A packing list such as this must be used by the material controller to check incoming receivables.

The material controller can use several different methods for checking quantity, each method being a perfectly valid procedure:

- *Full count,,* where 100 percent of the material is counted.
- *Sample count*, where an amount equal to at least 10 percent of the shipment is counted.
- *Scale count*, usually used when a large number of small items must be counted. In this method, a given number of items are counted, usually 25, or more, then the entire shipment is weighed and divided by the weight of the 25 units. This number is multiplied by 25 to find the total number of units in the shipment.
- *Theoretical weights and measures* are used when large, bulk items are received—such as steel, lumber, gasoline, oil, and the like—and it is simply not feasible to verify the packing list count or measurement. Therefore, cubic dimensions or volume are estimated, or the truck scale weight is accepted as verification.

10. When the count is completed, the material is tagged by the material controller with a move tag. (See FIG. 12-6.) The packing list is given to the receiving clerk, who records it in the computer file and sends the hard copy to accounting. In the event that the packing list for the shipment cannot be found—sometimes it is rubbed off the material it is attached to or it is buried in one of the cartons—then the material controller must prepare one, such as shown in FIG. 12-7.

11. When the receiving office assigns the material controller to a specific truck, he or she is given a printout that starts the receiving report in motion. At the same time, the inspection department is given a copy of this report form. Thus, as soon as the inspector assigned to this shipment is able to release the shipment, material control can transport the material to its properly designated location. This information will appear on the receiving report and will be marked on the move tag (FIG. 12-6) by the material controller.

12. An important exception to the orderly routing of the receiving function is the handling of "hot" items. If any part of the shipment is on the Short List, then the receiving report will indicate this fact, and both the material controller and the inspector will give such material top priority. After the material is inspected, the material controller will put a red-colored move tag, or some appropriate indicator, on the item and alert material control that the material is ready to be moved.

13. During the inspection of the item, if the tolerances of defectives are more than allowable according to company standards, then the material controller will set the material into a holding area after tagging it with a *hold for disposition* notice. The inspector then notifies purchasing and accounts payable so that purchasing can take action and the accounting department will withhold payment until such time as a negotiated settlement is achieved between the vendor and the purchasing department.

14. When the company is fairly large, it might have more than one manufacturing plant or warehousing entity, to the extent that materials might be shipped be-

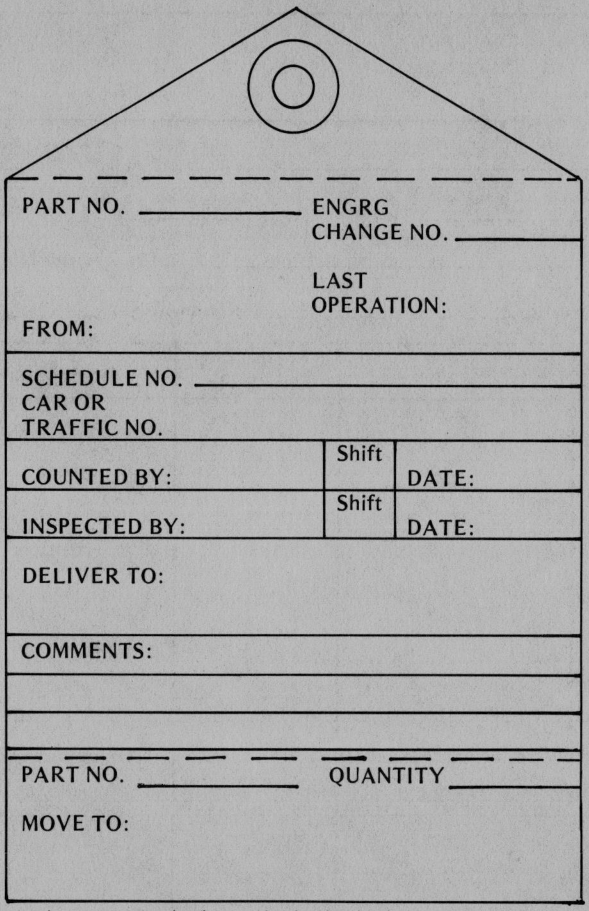

PART NO. _____ ENGRG
CHANGE NO. _____

LAST
OPERATION:
FROM:

SCHEDULE NO. _____
CAR OR
TRAFFIC NO.

COUNTED BY: | Shift | DATE:

INSPECTED BY: | Shift | DATE:

DELIVER TO:

COMMENTS:

PART NO. _____ QUANTITY _____

MOVE TO:

Fig. 12-6. An example of a Move Tag, which can be wire-tied or taped to the material. It is made from heavy, 80# card stock, and is attached by the material controller assigned to the receiving department.

tween plants. In such an instance, when an interplant shipment is received, the shipment is treated the same as any vendor receivable, with the possible exception that inspection is, generally, not required. The interplant shipment does not require a bill of lading, and it is sufficient to use an informal intracompany packing list for checking the material count.

Any discrepancies noted in the shipment of materials should be noted on the packing list. The material controller forwards the annotated packing list to the inventory control group for record-balancing purposes. Any discrepancies uncovered by the material controller should be resolved in a few days; usually a week's time should be sufficient.

Using automatic identification technology   The materials-handling procedures just discussed stressed manual counting. Outside of the computer interface to log

Pieces	Part No.	Engineering Change No.	Description	Weight

PACKING LIST   Form No. 12345

XYZ COMPANY
     CONSIGNOR _____

Terms _____ F.o.B. Point _____

Receiving Date _____ Invoice Date _____ Invoice No. _____

Shipped Via _____ Traffic Register No. _____

Purchase Order No. _____ Total Weight _____

Fig. 12-7. A substitute packing list prepared by Material Control.

information into the system, obtain receiving reports, and transmit data to other departments, there are many disadvantages accruing to the method of counting by humans unassisted by electromechanical means. Archaic methods such as these introduce many errors into inventory records that can be largely eliminated by automatic identification methods. For this reason, it is useful for the materials manager to have an understanding of bar coding and other auto-ID techniques. The control of inventory throughout the distribution cycle from raw material source, through manufacturing, storage, and the ultimate consumer will be positive and free from the counting errors that are anathema to the inventory department.

As an example, bar codes can be read by the use of pen wand scanners and laser scanners with excellent repetitive results. According to studies conducted by the Department of Defense, first-read error rates are in the neighborhood of one error for several million characters read, in contrast to human readable and transcription data entry of from one to three errors per hundred characters.

The bar coding symbols used in the supermarket and grocery chains are known as the Uniform Product Code (UPC). In the auto-identification craft, the other name for the UPC code is the "2 of 5" bar code. All bar codes, of which there are many varieties—over 40 different ones by last count—are composed of narrow bars, narrow spaces, wide bars, and wide spaces. If the code is printed horizontally, it is called a *picket fence*; if printed vertically, it is called a *ladder*, which is pretty much a descriptive type of nomenclature.

In order for the materials manager to obtain a perspective of bar coding, an encapsulated history of bar coding is given here:

1949   A patent for a circular bar code was filed by N.J. Woodland et al.

1960   Rail car identification symbology was developed by Sylvania.

1963   Bar code techniques were described in *Control Engineering*, October 1963.

1970   U.S. Supermarket Ad Hoc Committee on Universal Product Coding was formed.

1971   Plessey code was used in European libraries.

1972   Codabar patent was filed. The Interleaved 2 of 5 was proposed.

1973   UPC symbol was adopted.

1974   CODE 39 was developed.

1977   EAN (European Article Numbering) symbol was adopted. American Blood Commission Standard became Codabar. Code 11 developed, and the Department of Defense LOGMARS study effort was started.

1981   Distribution Symbol Study Group was originally commissioned by the Uniform Product Code Council to study the feasibility of printing bar codes directly on corrugated fiberboard cartons. Code 39 was recommended.

1982   Department of Defense, MIL-STD-1189, "Standard Symbology for Marking Unit Packs, Outer Container and Selected Documents" was published. (LOGMARS Program of the DOD is the acronym for Logistics Applications for Marking and Reading Symbols.)

The necessity for compiling and manipulating data quickly and accurately has been the driving force that has promoted the ubiquitous use of bar codes. In the receiving procedures that were just discussed, all of the manual counting can be accomplished by using pen-wand scanners, or in more mechanized receiving operations by using a fixed-beam laser scanner mounted alongside each receiving conveyor. It is possible by this means—i.e., using either wand or scanner—to trigger the following events based upon the bar-coded item:

• Automatic production of a bar-coded label that contains the traffic number assigned to the shipment in both bar-coded symbolism and in human readable characters.

- Automatic production of a receiving report based upon a summation of all bar-coded items received.
- Automatic data transmission to accounting of receiving data.
- Automatic comparison with a bar-coded freight bill, or manifest
- Automatic production of an Over, Short, or Damage report, as required
- Automatic production of a move tag.
- Automatic comparison with the "Hot" list and of a hot line move tag, as well as data transmission to material control that the wanted material has arrived.

In addition to these advantages, the bar-coded material and packing list can virtually eliminate "over and short" out of the system, except that which is due to pilferage. However, even pilferage is minimized because, with the stricter controls provided by bar coding, collusion between the trucker and the company employees becomes infinitely more difficult. Also, in interplant shipments, if the shipping plant is using bar coding as well as the receiving plant, all material counts should be accurate, thus eliminating the discrepancies of inventory levels.

Materials management must promote the use of bar codes for supplier purchases in order to streamline both receiving and tracking activities. Many suppliers are now using UPC, "2 of 5" code and CODE 39, as well as the "Interleaved 2 of 5" code adapted by parts of the automotive industry. Inasmuch as suppliers' use of a standard code is concerned, it depends to a large extent on what segment of the industry gives them the larger share of the market. The manufacturers of bar-coding equipment have largely solved this problem because most scanners made today are programmed to recognize the particular coded system they are scanning.

The material management department, therefore, must attempt to standardize on the bar-code label system it finds most desirable. However, the department also should be aware of the different codes it might be forced to receive from certain large suppliers, who might be relatively inflexible in their own auto-ID label making and use scanning software capable of adjusting to a number of the more popular bar codes.

Sometimes the inflexibility of bar-code determination rests not with the supplier, so much as it does with the customer. As an example, I was a member of the joint Department of Defense and General Services Administration LOGMARS Committee, which established the CODE 39, or "3 of 9" bar code for military logistics. This culminated in the promulgation of MIL – STD – 1189, which is a description of the code structure. Since both DOD and GSA buy billions of dollars of supplies annually, the "3 of 9" code is probably one of the most widely used codes, in addition to the UPC, or so-called "grocery" code.

The "3 of 9," or CODE 39, bar code was developed by Dr. Allais, then president of the Intermec Corporation of Lynnwood, Washington, in 1974. It is a fully alphanumeric bar code system with a structure that enables it to be printed by a number of printing methods, among them offset, letterpress, impact printers, dot matrix, and nonimpact printing devices. The "3 of 9" code can be used for inventory control, tracking of manufacturing work in process, wholesale and retail distribution, property accountability, automotive parts production and assembly, and airplane assembly and maintenance, to mention only a few of the areas that it can serve. (See appendix A for a list of automatic identification and peripheral equipment suppliers.)

Production work orders   There is a very close relationship between material control and production control. In fact, if there can be said to be a lexicography of manufacturing, the phrases are virtually synonymous. The differences appear to be largely that in the material control area the emphasis is on the movement of materials, and in production control the movement consists, in the main, of moving the paperwork or data transmission that makes the movement of materials a logical outcome of this endeavor. For this reason, the combination of data transmission and material transportation has a unified appearance and marches along under the heading of *production control.*

In the area of production control, then, the production work order is the instrument that gets things done, for it is the document that authorizes the machining or assembly of materials. In a nonmanufacturing context, for example, a hospital, the production work order might very easily be considered the document that authorizes blood tests, CAT scanning, or the like. Nevertheless, in the realm of data transmission, the purpose of the production work order is to supply essential information to the manufacturing machine shops and the related areas of accounting and inventory control.

The accounting department uses the information concerning the first and last operations as its authority to transfer inventory quantities from rough stores (raw materials) to work-in-process, and ultimately to *finished stores* (completed work resulting in end products, be they the actual end product or spare parts and components that are to proceed to the parts warehouse or distribution network). The accounting department accomplishes this task by transferring half of the burden and labor costs along with the rough to in-process data upon notification of the completion of the first operation to be performed on the part from factory (or plant) accounting to the general accounting office. The in-process inventory is then transferred to the finished status, along with the remaining half of the burden and costs after notification by manufacturing that the last operation on the part has been completed. Each transmission of first and last operation data sent on its way to general accounting by the factory accounting personnel gives the part number and the quantity of pieces that is to be transferred from one inventory status to another.

The production work order is the means by which inventory control can track worked material. Thus, when a specific work order has been completed, the inventory control stock record reduces the rough on-hand balance and increases the finished on-hand quantity so that a balance is achieved. When this up-dating of the record is finalized, the specific work order schedule is deemed complete, and the computer file will bear this close-out notation.

As just indicated, the manufacturing department uses the production work order as its notification to begin production on the part. It lets the shops know what to produce, when to produce it, and in what quantity.

The materials handling in each of the manufacturing processing shops is performed by a dedicated material control crew. Therefore, as the work orders are generated, each of the assigned material control groups receives the work order for their area. If RF receivers and terminals are mounted on their forklift trucks, then each material controller can receive instructions individually for his or her next movement of materials. By keyboarding the terminal console, the operator can signal the start and finish of his or her operation, receive the proper instruction based upon an employee number, and get credit for the material movement. In this latter instance, the employee number, keyed together with the part number, indicates both a method for tracking material and a means to collect work measurement data.

In lieu of the operator punching in his or her employee number, a magnetic stripe card or a bar-code label that can be pen-wanded at the time of the pickup of the material will serve the same purpose. Also, at each workstation, the machine tool operator or processor can indicate in a conveniently located terminal by means of his or her badge number the starting and ending time for each operation, which also helps sort out the whereabouts of each piece part, supplies data to the industrial engineering time study group for verifying work measurement standards, and permits proper labor-costing to be accomplished.

Standardizing the data transmission method for material controllers is relatively important, despite the fact that most of the numbers may stand alone. Nevertheless, it pays to have a checklist of the order in which data is to be transmitted because it helps the operator routinize the procedure, as follows:

- Employee badge number
- Part number, plus engineering change number, if any
- Name of part
- Schedule number
- Lot size
- Location of first and last operation
- Part number from which this part has been made, if applicable
- Shop area number
- Next delivery destination, or production line location
- Date work order received

Inasmuch as material management activities have been viewed from a total systems standpoint in this text, it will be necessary for the material control group to establish routines that will take into consideration the several exceptions to the smooth and regular flow of materials through the production process. In the first instance is the method for handling scrapped parts on either the production lines or in final assembly when the end product, for one reason or another, must be scrapped. Adjustments in the inventory records must be made, and the necessary procedures must be established to handle rework and salvage operations.

Another difficulty that often crops up in manufacturing plants is the necessity for changing quantities on work orders after the first operations have been initiated. These changes could be in time—i.e., date required—location for machining, machine process, quantity, and so forth. Sometimes, the entire work order is canceled.

In the case of a cancellation, which is usually originated by the inventory control group, the primary objective is to prevent the part from entering the production process. If this is not possible because processing has been started, then the task is to stop production before too many pieces have been run, and then to properly dispose of or salvage all these parts, if possible. All of the changes made by the material controller and reported to his or her supervisor must be coordinated with inventory control and all the workstations that are a part of the original production work order so that all accounts balance.

When material that has been completed and has been recorded as such to the inventory control group is returned to the shop for rework, it should be identified correctly by the materials controller as rework and not added a second time to the inventory group as a completed piece part, since that would inflate the inventory records. Therefore, the proper identification of the material as rework is the key to maintaining accurate inventory records.

Keeping production lines supplied   An important aspect of the materials-handling function performed by the material control group is the task associated with keeping the production and assembly lines supplied with materials. In Just-in-Time manufacturing, this means a day's supply to several day's supply at each workstation. In most American plants, however, this supply quantity may vary from a few hours' worth on hand (very rare except in some alley shop operations) to several months', with an average of approximately a week's supply at each station.

A material controller assigned to assembly and production line supply will learn to eyeball the consumption of stock on his or her line and make judgments as to quantities of materials required. In addition to visual inspection of each workstation, however, the material control handler should maintain a checklist. Depending upon how computerized the manufacturing plant is, he/she will have the following information in hard copy or on a computer terminal screen:

- Part number and name (description)
- Model or assembly on which it is used
- Quantity required per model
- Storage area location according to bin or floor area
- Source of supply (when applicable)
- Quantity on hand (updated daily in a manual system of recordation, or continuously updated in a real-time computerized system—or batch updated)

In addition to the checklist there is a Master List, which is usually subject to frequent revisions as engineering changes are made. In a manual system, the material controller or his or her supervisor has the responsibility for updating the master list; however, in a computerized system this is usually done in the Scheduling Section via data transmission to the Material Control Section, and thus is available to the material controllers assigned to the task of replenishing the line.

In addition to the activities of supplying the line, the material controller should replace any container tags that have been damaged or torn, or become illegible. He or she should also assist the workstation operator and the person assigned to the custodial chores in maintaining an orderly appearance in the shop area.

The material controller should review every piece part on his or her production line daily so that a five-days' supply of materials will be available for each operator. With some materials, this task might require a piece count to determine how close the operator is arriving at a depletion point. If the material has no backup in the storage area, then the material controller must develop a sense of urgency and initiate a "material short" document. This has to be done immediately because it indicates a breakdown of the system.

A manufacturing plant should never have to operate on a "short sheet" basis. If the proper planning, processing, purchasing, and material control functions have been properly coordinated, running out of materials or parts should be virtually unheard of. Nevertheless, a short slip should contain the following information:

- Part number and description
- Date of the action
- Quantity on hand
- Usage per day
- Sourcing area
- Location where required

The material control handler should use a special routine to transmit the information to the material control clerk at the location where the needed part is produced, assuming that the part is to be made in-house. If it is a purchased part, then the purchasing department must expedite the requisitioning and delivery of the part. Simultaneously with the discovery of the part's short supply, the receiving department and inventory control are alerted to the emergency and a hot-line network is established.

In a fully computerized manufacturing plant, the response to a stock-out emergency can be handled in an almost routine manner; however, a sense of urgency must be inculcated in all of the employees that the emergency is indeed serious, since stopping the production line might lead to layoffs, lost contracts, and other penalties. Job security and self-interest can be powerful motivators, and the very essence of good shop floor control means that the production and assembly lines must be stocked adequately at all times.

When a stock-out of a part occurs in a manual system, the supervisory material control person should write a full report that is sent to the plant manager. It is up to the plant manager or his or her designee to investigate all such shortages and institute revisions to the procedures and personnel wherever they are required. Obviously, if the system is without blame, then the shortage was due to a personnel delinquency, and staff retraining, etc., might be required.

In either a manual or computerized control system, the material control clerk (or the scheduling section) should check the schedules to see if there is an open work order for the required part that has a sufficient balance to cover the shortage in question. If there is such an open work order, the schedule number should either be placed on the short slip or entered into the computer in the proper routine. If there are no work orders open or if work order schedules are insufficient to satisfy the shortages in addition to balances required through the work order period, then the inventory control group must initiate a work order. Inventory control will notify the material control group of the work order schedule number and the lot size they are able to release. This information will be entered by the material control clerk on the short slip or into the computer file, as the case may be.

If the manufacturing department has more than one item in short supply, then all of the short slips will be tabulated into a "short sheet," and transmitted by material con-

trol to all of the applicable shop areas. It is an unfortunate state of affairs in any manufacturing or service industry when a short sheet exists, and the onus rests not only with
the plant manager but with the top management of a company that permits such a dis-
advantageous problem to persist.

For the sake of the previous discussion, purchased finished or interplant materials
that are found to be in short supply should be treated with the same sense of urgency
and with a systemized approach that will allow for the expeditious handling of any part
in short supply.

Handling scrap and surplus materials   In most organizations, scrap and
surplus material is separated one from the other just as it is in an accounting sense.
Parts that have to be scrapped because they are defective present no really large problem except in two areas. One is a question of material. For example, steel and iron scrap
can be sold to scrap dealers and some small monetary return is possible. Materials such
as plastic might have a market as scrap, but it depends largely on the type of plastic
involved. There is a small monetary return that can be recaptured with this material, as
well.

The other area for concern is that when scrap is generated on machine lines, it
should be monitored very closely by manufacturing since reworking the part might add
additional manufacturing cost and can upset schedules. Where the part cannot be salvaged, the lost hours, wear and tear on equipment, and cost of handling and storage all
subtract dollars from profitability. The plant management must decide what it con-
siders to be acceptable scrap levels—for an inevitable amount of scrap is generated in
any manufacturing enterprise, regardless of how well run it appears to be—and then it
must take appropriate action to improve the situation, whether it be to purchase newer
and better machine tools, retrain workers, etc.

When a part can no longer be used because it has been made obsolete by an engineering design change, it cannot be reworked to bring it up to the new specifications,
current sales volume can no longer justify keeping the part, or a number of other reasons including overproduction, then the part must be declared surplus. The inventory
control group should be entrusted with the responsibility for originating a request to
scrap surplus material. Therefore, they must request that a count be made, and a material movement notice must be initiated. The data contained on this notice should be
transmitted either manually or electronically to all production and assembly lines, as
well as all storage locations where the part was either used or stored. This notice
becomes the activator for the line material controller to move the material to the surplus
material location area of the plant, where this material will be stored and counted. The
results of the count, together with the data, shown in FIG. 12-8, are transmitted to inventory control.

After the material has been moved and counted, the inventory control group is
responsible for initiating a document that is, in fact, a request to scrap surplus material.
(See FIG. 12-9.) The surplus material is released for scrap action after the salvage committee reviews the history of the material and signs off on the "request" form, and the
plant manager countersigns it.

Although it is possible to avoid the cumbersome paperwork involved in scrapping
surplus material, the condition or action is such a serious and costly event that the

SURPLUS MATERIAL COUNT AND MOVE DATA

Part Name ____   Code ____   Part No. ____   Date ____

Status	Physical Count	Inventory Control Record	Shop Location	Move to Location	Status of Open Work Orders				Parts & Interplant Orders				Stores Record	
					W.O. No.	Amount	Delivered	Finished	Assembly No.	Quantity	Date	Bal. Due.	Date	P.O. No.
Rough														
Semi Finished														
Finished														
Finished													Stores Count Amt.	Sign.
Finished														
Total														

Record Total After Cnt	Rough	Finished	Remarks

Status	Adjustments		Request No.
	Plus	Minus	
Rough			Adjustment approved by:
Finished			Adjusted by

Adjustment approved by
Inventory Control Manager ____

Production Manager ____

Counted by: ____   Date ____

Count requested by ____

Reasons for declaring surplus

Obsoleted Part:
Repl. by ____
Super. by ____

Current Part:
Scrap Engr. No. ____
Surplus 2 years ____
Other: ____

Fig. 12-8. Data for counting and moving surplus material to the scrap location within the plant, or in the plant complex.

REQUEST TO SCRAP SURPLUS MATERIAL AUTHORITY				No.		
Description		Model	Material Code		Part No.	
Current Part Scrap per Change No.	New Part No.		Change No.		Class	
Non-current	Cancelled		Surplus Code		Worked Purch.Fin.	
Approved by	Surplus Quantity			Location of Material		
	Rough	Semi-Fin	Finished			

Standard Costs		
Unit	Total	
Unfinished Material	Rough	
Purchased Finished	Semi-Finished	
Hours	Finished	
Total	Scrap/Surplus Total	
Total Std Costs		

Reviewed by:	Date	Authority is given to scrap	
Parts Dept.		Rough	
Inv. Contr. Supvr.		Semi-Fin.	
Chief Engr.		Finished	
Purchasing		Approved by Salvage Committee	
			Chmn.
Issued by	Date	Recorded by	Date	Plant Manager

Fig. 12-9. An example of a form that can be used to declare surplus material scrapped.

paperwork is used to denote the gravity of the judgmental error that permits the situation to exist. An electronic transmission is all that is needed in a computerized manufacturing system to affect the transfer of this material; however, because it sometimes represents large sums of company money, it is advisable to treat this action seriously, if only to impress on all the departments concerned that collectively and individually they bear the responsibility of minimizing, or maximizing, profitability.

The material management's material control group has the largest share in the actual processing and handling of surplus material that is to be scrapped. As shown in FIGS. 12-8 and 12-9, there are a number of plant departments that must review the requests for moving surplus from the production and assembly lines to the scrap location.

The last department of the plant that might have a use for the item is the parts department. This department is the last repository in the company for obsolete parts. One of the chief reasons is that obsolete parts are often required by the parts department to fill customer requests.

The product's aftermarket is often as lucrative as the initial product sale, and sometimes more so when the parts are marked up several hundred percent. Although this markup seems exorbitant, consider the company's per-annum storage costs, which often amount to 25 percent of the item's cost of production. Some companies keep replacement parts for 10 or more years after the end product is out of production. Other companies, who pride themselves on service, will produce spare parts from blueprints when they are out of stock on an item. It might take a little longer to obtain the part and the cost might seem unreasonable compared to a new product part, but since the old part might result in placing the entire machine back into service, most customers of big-ticket items prefer to pay the somewhat exorbitant spares replacement costs, rather than buy a new model machine at today's inflated prices.

Thus, the parts department will review the demand history for the part and might decide to request that some, if not all, of the surplus items be sent to their warehouse.

Another review of obsolete parts is actually accomplished by the parts department. It will canvas its dealer network to sell, or consign, obsolete parts to dealers who might be servicing some of the older machines and might wish to replenish their stores of these items. When the parts are consigned, the dealer will pay for a percentage of the cost initially upon receipt of the item, and the balance of the cost when the part is finally sold to a customer.

Since every effort should be made to prevent a part from being relegated to the scrap pile, parts that are surplus but are unacceptable to the parts department because they are dirty, rusty, require paint, etc., and so forth should be refurbished with equipment and supplies that have been provided for this chore. Additionally, specific personnel of the material control group should be assigned to this task on an "as required" basis.

Manufacturing materials handling   Although the previous discussion of one of the more important areas of materials handling has involved a concentrated focus on the line material control activity, there is an underlying concept concerning the flow and movement of material that must not be neglected by the materials management organization. It has to do with optimizing this flow of materials through the plant. The plant in this instance could be a warehouse, factory, military installation, brokerage house, hospital, or any other kind of manufacturing or nonmanufacturing entity. The fact remains that they all have a common requirement: in order to operate most effectively they must have an orderly flow of materials. It does not make any difference if it is a question of food to be prepared, orders to be filled, or parts to be made; there is a requirement to provide an orderly flow from the beginning to the end of the distribution chain—i.e., from the source of raw materials to the ultimate consumer.

As an example of orderly flow in materials handling, you need only look at the prevalence of back-hauling in many companies. *Back-hauling* simply means that instead of maintaining a direction, material retraces its path unnecessarily because of poor layout and poor organization of the materials-handling process. In addition to adding to the handling time, back-hauling increases the wear and tear on equipment, and adds to the cost of the handling function. It is so widespread as a common fault of handling that the materials management department must be constantly vigilant to review handling operations, especially where new production or assembly lines are being planned. Sometimes it is necessary to review material control functions with this thought in mind; however, there are many areas in both manufacturing and nonmanufacturing organizations in which a periodic materials-handling survey will achieve beneficial results because there is more to the optimization of materials-handling activities than the elimination of back-hauling.

As an example of general awareness concerning materials-handling methods, materials management should ask the following questions in each material control area:

- Does every aisle permit the easy passage of materials handling (MH) equipment?
- Have any aisles in the area been blocked so that free passage is not possible?
- Have all of the aisles been properly marked where required by OSHA?[1]
- Are all MH vechicles being operated at safe speeds?
- Can MH equipment safely pass each other at intersecting aisles?
- Is pedestrian traffic endangered by MH equipment operations?
- Is the plant and area housekeeping up to good operating practices?
- Is the condition of the plant floor up to standard in the following areas: working aisles, main aisles, and around machinery?

From these questions it is easy to determine that there is a direct relationship between the way in which work is performed—i.e., its orderliness—and the flow, or movement, of materials through the plant. In addition, with the advent of the OSHA regulations, which have spurred state legislatures to adopt similar laws, there has been an impact on the safety aspects of plant operation that implies that good housekeeping is important and a necessary part of management's task.

Some of the OSHA requirements stipulate minimum aisle dimensions, safe in-plant vehicular speeds, and aisles wide enough to accommodate both vehicular and pedestrian traffic. Some aisles in your plant, for example, might be wide enough for two oncoming vehicles to pass each other, but they must also have the necessary additional 3 feet of width for pedestrian traffic. In addition, aisles and pedestrian walkways must be marked clearly in order to prevent accidents.

Floor conditions are also important safety factors and have a great deal to do with making material movement effective. As an example, oil spills, shot-blast pellets on the floor, loose wood blocks, or spalling concrete not only contribute to high accident rates, but tend to reduce productivity and increase product damage, and are often responsible for low employee morale. Who likes to work under conditions such as these? The worker gets the impression that management doesn't care, so why should he or she?

In manufacturing materials handling, the materials manager should coordinate the efforts of his or her material control group with the manufacturing materials-handling engineer. There are many ways that materials handling can be improved in manufacturing. Some of these areas include the following:

- Reduce the amount of materials handling being performed by direct labor. In other words, the higher-priced machine operator should not have to expend time or energy obtaining parts or materials.
- Keep an adequate supply of materials at the workplace. Do not have operators standing around waiting for materials or tools.
- Have materials delivered to the operator's workstation without interrupting the work cycle.
- Make available, insofar as possible, mechanical loading and unloading devices for each machine tool that requires such devices. Ascertain whether or not these devices are being put to good use and are being used properly.
- Remove waste and trash from the workplace without interfering with the work cycle.
- Provide sufficient in-process storage area, but not so much as to be excessive. Remember manufacturing floor area is very expensive.
- Determine whether or not the methods employed to load and unload the following systems are adequate:
  - washing and cleaning
  - painting
  - heat treating
  - shot blast, or shot peening
  - other systems
- Have materials readily accessible to assemblers and machine operators, i.e., design workstation arrangements for easy access of materials and convenience of the operators. Use physiological factors to adapt the workplace to the operator, remembering at all times that humans differ in height, reach, and strength.
- Wherever it is possible to gain a manufacturing productivity or handling advantage, use racks for the storage of materials on assembly lines or in the assembly area.
- Make use of air rights (and cubic volume of plant), rather than area alone, in compressing the workplace for the operator's advantage, or the materials-handling advantage that might be gained thereby.
- Remember to match containers with the materials they are transporting and to design containers so that they are functional and provide the maximum security for the product and convenience for the operator.
- Help keep line shortages to an absolute minimum.

In the cooperative effort between manufacturing and materials management, it is hoped that many more decisions affecting the materials-handling effort will be made prior to the installation of complex manufacturing systems, rather than after the machines are operating. As an example, companies usually spend large amounts of capital on such elaborate systems as washing and heat-treating machines and spend relatively less thought on the manner in which materials are loaded and unloaded from the machines, although in a systems approach to materials handling this aspect of the operation would not be neglected. For this reason, carriers on paint lines might sometimes be designed to pick up and discharge the part automatically. The product might even be brought to the pickup point by conveyor and removed from the paint line in similar fashion. Also, some types of heat-treating furnaces might be loaded and unloaded by means of tub dumpers and chutes.

The means for mechanizing some of these systems should be considered prior to the purchase of the equipment, since the decision to convert or modify systems that have already been installed is usually more costly than if made before the system has been fabricated and placed on the site.

It has been mentioned that materials should be readily available to the operator. In this category should be placed the requirement for portable and expendable tools such as drill bits. Consideration should be given to the use of pneumatic tubes for tool delivery of these types of expendable tools. In addition to the means for tool delivery, there should be pneumatic tool-delivery stations at convenient locations within the production areas so that machine operators would have ready access to these tools.

Equipment and maintenance

Since the materials management department is keenly aware of the overall cost of operation, it will try wherever possible to economize and hold down expenses. One of the places in which it is necessary to spend more to earn more is in equipment maintenance, for when materials handling equipment is poorly and inadequately maintained or when the equipment is overworked, maintenance costs also increase. Contrary to expectations, equipment costs will rise when too much equipment—that is, underutilized equipment—is on the books (capitalized by accounting).

In order to determine how much it is really costing the company to maintain and operate its mobile materials-handling equipment, there must first be a comprehensive program of preventive maintenance and record-keeping that is accurate and consistently applied. Record-keeping, by and of itself, however, will not tell the true story of what is happening to operating equipment without constant analysis of maintenance records and an action program to make the results of the analysis effective.

In the first place, the functional responsibility for mobile materials-handling equipment maintenance should rest within the materials-handling group. Therefore, the control and allocation of equipment throughout the plant should be included in this group's functional statement. In some companies, functional responsibilities are often divided among several groups, leading to a certain redundancy between operations and operators that contributes to a higher cost of manufacturing and distribution.

Also, mobile equipment and automotive maintenance sometimes are placed under the aegis of plant engineering. In this situation, if there is no formal structure to preventive maintenance and no schedules are in effect, what usually happens is that the departments that complain the loudest often have their equipment serviced, whereas equipment maintenance for other less vocal departments is very often sidelined in preference to the more vociferous group. Some departments have more clout than others, and this leads to other preferential treatment, sometimes at a disadvantage in cost or schedules.

Even when maintenance costs creep up higher and higher, as they do in some companies, the maintenance department rarely has trouble in obtaining as much additional budgetary consideration as do other plant departments, simply because no one seems to take the time to analyze maintenance costs properly.

To determine whether or not your company has an effective materials-handling equipment maintenance program, it would be advisable for the materials management organization to conduct a survey. The most important question of the survey is: Does this facility have and maintain an up-to-date MH equipment roster? If the answer to this question is yes, then does the list contain the following data:

- Equipment number—each piece of equipment should have its own ID number.
- Name of the manufacturer.
- Year the equipment was purchased. (It is surprising that some companies have to spend time looking through past invoices to find the answer to this question; however, in an intelligent MH equipment program, this information should be on an equipment card, and in the computer data file.)
- Equipment specification. (This information should also be on the equipment card and in the computer data file.)
- Area of the plant to which the equipment has been assigned. (If the piece of equipment is not in the area to which it was originally assigned—i.e., if it has been traded, lent to, or borrowed from another department—then you know that your MH equipment program is in trouble.)
- Probable replacement date. (This is an added bit of sophistication in your MH equipment program because it involves a great deal of analysis, even in a well-established MH equipment maintenance program.)
- Initial cost.
- Attachments. (Sometimes attachments have been added to mobile MH equipment. In this case, the pertinent data just listed—i.e., manufacturer's name, year purchased, specification, and cost—should be shown on the equipment card and entered into the computer data file.)
- Total dollar value of the fleet. (This value changes from accounting period to period as equipment is added, dropped, and depreciated.)

In addition to this data for each piece of equipment, there are other areas that are worthy of surveillance in the logic of the systems approach to materials management, of which the mobile MH equipment maintenance program is but a phase. For example, it would be appropriate to determine just how precise maintenance cost records are, and how effectively cost control is practiced.

The cost per hour of operation for each vehicle is one of the important concerns of any good MH equipment program. In this regard there are alternative methods at each company's disposal: to buy or lease the equipment, and to perform the maintenance in-house or to obtain contract maintenance. In-house maintenance would make use of the regular plant maintenance workforce, but contractual maintenance would be for services at a guaranteed rate, performed at either the contractor's premises (which would free up valuable plant space and should be considered in any R.O.I. evaluation as a trade-off) or in-house with the contractor's own personnel.

In-house maintenance usually has the capability of providing fast service because it is close to the scene of operations. However, although this method might give the appearance and feeling of security it is not necessarily the case because it might just mean that the company is doing a lot of unnecessary firefighting that could be avoided if a well-planned maintenance program were in effect at the company. Thus, the company's MH maintenance program requires a systems approach, combining such elements as:

- Routine inspection, servicing, and lubrication.
- Periodic tune-ups.
- Performance of minor repairs and replacement of parts before breakdowns occur, and the like.
- Major overhauls at prescribed frequencies.

In comparing in-house work-force maintenance with contract service plans, the materials management organization should evaluate performance of the alternative methods in these four areas mentioned, as shown in TABLE 12-1.

Table 12-1. Comparison of Maintenance Alternatives.

Area	In-house	Contract Service
Routine service	Satisfactory only when schedule is properly implemented.	Specializes in this type of performance.
Periodic tune-ups	Satisfactory only if company mechanics have the proper skills.	Specializes in this type of performance.
Minor repairs	Usually satisfactory if done on time; however, competes with other maintenance chores and problems	Unless this service is performed in the client's plant, would have to be called when needed.
Major overhauls	Competes with other maintenance jobs and other uses of maintenance facilities; downtime can become a problem, unless units are leased immediately.	Replacement units from the service company's rental fleet can be provided during overhauls, etc., as a provision of the service contract, without incurring excessive downtime.

In general, the service contract approach meets the company reliability requirements more completely; however, the in-house type of maintenance program might be just as effective and less costly under certain conditions. Some of the pros and cons have been delineated, as follows:

- *Qualified personnel and suitable facilities* Larger companies have the wherewithal to provide in-house maintenance service that is on a par with contractual service, since they can afford adequate facilities, equipment, and trained mechanics for all the plant's mobile MH vehicles. The possible exception would be large cranes and specialized lifting equipment, such as magnetic lifters. Usually, the in-house staff services only MH equipment, and other maintenance chores are taken over by plant engineering.

In smaller companies, regular vehicular mechanics service MH equipment along with all of their other rolling stock, and the mechanics are not often familiar with this type of equipment unless they have received factory training. Also, in some smaller companies, the special tools and parts are not always immediately available to perform even normal MH equipment maintenance tasks, so the chances are strong that most of the MH fleet will get short shrift unless a vehicle breaks down completely, develops a bad oil leak, or has a slipping clutch or bad brakes, or if lift cylinders go soft.

Another disadvantage of in-house maintenance, especially in smaller companies, is the competence of supervision. Good supervisors usually are able to obtain and retain competent mechanics in the maintenance department, and when this combination exists the reliability of the maintenance performed establishes the reputation of the department. The plant's management, in this instance the materials management department, would not venture into the alternative of contract maintenance with a good in-house program already underway.

On the other hand, if the company does not possess a qualified maintenance group, it would look outside the company to establish a maintenance contract. However, the materials management task would be to ascertain, or have certified, the proficiency of the contractor's mechanics and track record, before negotiating the matter of price.

- *Keeping spare parts on hand* Having enough spare parts, or the right parts, for MH equipment maintenance is a problem even for large multinational companies. Standardization of equipment will reduce the severity of the problem, but since most equipment fleets grow like Topsy, with different models even from the same manufacturer, even companies that do an excellent job in equipment standardization occasionally have a problem obtaining parts. The large, or small, company with a mixed fleet always will have a parts problem. However, sources of supply are getting very responsive to these concerns, and the problems might not be as serious as they once were, depending on the proximity of the plant to large metropolitan centers.

  The materials manager must consider the spare parts supply situation when considering the alternative methods of maintenance service programs. For this reason, contract maintenance service organizations must resolve the parts problem prior to obtaining, and as a condition of, a service contract. Thus, before the company is committed to a service contract, the quality of parts supply should be researched exhaustively to determine how well the contractor is performing with its existing clients.

- *Reliability considerations* Although it is valid to assume that larger plants might be better able to provide a satisfactory maintenance program for MH equipment, they do need to consider maintenance contracts under certain circumstances, especially when equipment availability or uptime is the criterion. For example, it might be more economical to use a service contractor than to perform in-house maintenance, especially in areas where labor rates are high and fringe benefits amount to a substantial amount of additional cost.

  Thus, if the service contractor can guarantee a specific level of maintenance service, their proposal might be very advantageous for the company. If, however, there are other factors that enter into consideration such as that the service contractor might not be able to guarantee a certain level of equipment availability because of such circumstances as poor proficiency of equipment operators or an arduous operating environment—such as a foundry or steel mill, the average age of the equipment, poor operator training, poor plant supervision, and so forth—then the contractor will not not be able to give any guarantee whatsoever.

Whether to consider a service contractor or to provide a well-supported in-plant maintenance program is a decision that should be evaluated in terms of the satisfactory provision of all elements of a maintenance program. Therefore, in considering contract maintenance services the following checklist might be of help:

A checklist for considering maintenance service contracts

✔ Will the service requirements meet minimum needs or standards?

✔ Will the services ensure flexibility in meeting the environmental conditions in which the equipment must operate? (A steel mill or foundry as opposed to the relatively clean and dust-free atmosphere of a warehouse, for example.)

✔ Will the equipment be inspected on a periodic schedule such that equipment problems will be discovered before they result in costly downtime or equipment replacement?

✔ Will the service contractor maintain an adequate supply of parts so that unnecessary downtime will not be incurred?

✔ Will the contractor provide loaners—i.e., replacement units—as required?

✔ Are all of the contractor's clients satisfied with their programs?

✔ Will the contractor provide a total maintenance program? If so, how flexible is the program for handling unscheduled (unplanned) problems? Are premium charges due for work of this type?

✔ Are major overhauls and equipment rebuilds considered a part of the contract?

✔ In the event of major overhauls, etc., will replacement equipment be provided during these long periods?

✔ Where will the contractor perform his or her service—in-plant or outside?

A checklist for considering in-house maintenance

✔ Is it possible for the regular maintenance work force to adequately support an MH maintenance program?

✔ Are the facilities and space for maintenance adequate? If they are not, will budgetary and other resources be made available at a later date? If they are not now available, can an adequate level of maintenance be provided during the interim?

✔ Are competent mechanics available? If they are not and factory training for certain equipment is necessary, will this training be made available to them?

✔ Do other maintenance chores need to be performed by this same maintenance crew? Is their group adequate in size to perform both equipment and plant maintenance? What part of this burden can be shared by plant engineering? Can the maintenance crew be dedicated to MH equipment maintenance only?

⮑Is the mobile MH fleet a mixture of odd makes with unusual spare parts require-ments? Is any of the equipment too old or obsolete?

⮑Can the in-house mechanics, as well as the purchasing department, deal with these problems in an adequate manner?

⮑Can the in-house maintenance group guarantee a specified level of equipment up-time?

If the materials management department cannot evolve a clear-cut decision con-cerning the alternative choices, then it might very well be that the final decision will be made based upon cost, as well as reliability.

Records and record-keeping

One of the major problems in evaluating the cost of maintaining materials-handling equipment is the availability of cost data. Almost every unit of mobile MH equipment comes equipped with hour meters, whose sole purpose is to indicate running, or up, time for the equipment. If hour meters are continuously vandalized, there is a discipli-nary problem involved that is very difficult to control without draconian measures. Sometimes it is possible to convince the shop stewards (in unionized plants) to assist in controlling the problem; however, whatever measures necessary to overcome this diffi-culty should be employed because without workable hour meters, a good maintenance program based upon scheduled service periods becomes next to impossible to achieve.

Hour meters are important tools in allocating equipment to the various plant departments based upon usage. Also, in developing replacement criteria, the operating hours of each piece of equipment become quite important.

Since MH equipment is crucial to the functioning of the manufacturing (and, in some instances, nonmanufacturing) plant, second only to production equipment, it also represents an important aspect of capital investment. Thus, its subsequent repair and maintenance will have a substantial economic impact on the MH equipment fleet and the company's profitability. In this regard, the cost analysis involving equipment should be based upon two considerations:

1. Repair and maintenance costs per unit of equipment
2. Total cost of fleet ownership and use

Each unit of equipment should have its own cost records, indicating on a monthly basis the amount of maintenance labor and materials expended. Monthly data should be summarized on a year-to-date and on a purchase-to-date cost basis.

Although such data helps single out the bad performers in the fleet, it does not tell the whole story, since consistently higher labor and material costs might indicate that planned—i.e., scheduled—component change-outs, rebuilds, overhauls, etc., might be missing from the maintenance program. Excessive material costs at irregular intervals might, in general, indicate that scheduled maintenance is not being accomplished, but has been replaced by break-down maintenance, or firefighting. This kind of mainte-nance is usually indicative of either poor supervision of the maintenance shop, under-staffing of the group, or no formal, scheduled maintenance program.

There is scarcely any reason today for not having a scheduled maintenance plan for the shop. A number of computer software programs are currently available for equipment maintenance. In addition, there are also a number of companies who have established themselves as equipment maintenance scheduling experts, who will come into your plant, survey the MH equipment fleet, take all of data required for each piece of equipment, and establish a scheduled maintenance program that will indicate when each piece of equipment must be serviced and inform the shop what maintenance service must be performed during the appropriate period. These companies will perform the same type of service function for production equipment, as well.

At least, with this approach commonly available, there is hardly any excuse for not having a well-thought-out and scientifically prepared program of scheduled maintenance in today's manufacturing establishment. The same systematic maintenance programs are available in the nonmanufacturing and service industries, also, and deserve consideration, especially where the planning time or the necessary paperwork skills are not available in a given plant situation.

Some of the information to be obtained from analyzing repair costs for specific MH equipment units is as follows:

- Widely divergent total costs for equipment of the same model and year should assist in singling out possible operator equipment abuse, which results in higher maintenance costs.
- Cost that differ widely for the same type and year of equipment used in the same operational area might point out a bad piece of equipment, i.e., a "lemon."
- Maintenance costs for a unit that are higher for each successive maintenance period might indicate the need for a major overhaul or for equipment replacement.

The data obtained from the analysis of labor and material costs can help establish the total cost of in-house maintenance. This information is useful when making a comparison of the cost of guaranteed contract maintenance services. If the materials management department has enough data on each piece of MH equipment, it is then possible to develop a broad estimate of the comparative value of contract services. The cost for each unit expressed on an annualized basis will help firm up the in-house vs. contract service price. If contractual services are obtained then the continuing collection of data for the maintenance cost of each piece of MH equipment will serve to determine the effectiveness of the purchased maintenance services during the contract period.

Since the materials management department is concerned with viewing its own operations on a broad, systemswide approach, it is necessary to integrate the MH equipment maintenance function into the larger perspective of plant operations. The total systems approach thus involves the economics of equipment ownership, which includes the following:

- Capital investment
- Depreciation
- Maintenance cost

- Cost of downtime
- Availability of equipment (uptime)
- Cost of equipment obsolescence

When all of these items are considered, then a true comparison for in-house vs. contract maintenance service can be attempted. These numbers also will provide a means for making decisions concerning each unit of the MH fleet and the management of the fleet as a whole.

Equipment replacement, justification, and R.O.I.

In most companies, regardless of size, the forklift truck is the universal MH equipment of choice. The selection of forklift equipment, then, can be categorized largely upon where the equipment will be used, the duty function of the equipment, and the size of the loads it must carry.

If speed of operation is an important consideration and the vehicle will be used mainly outdoors, then a gasoline-engine truck would get the first call, followed by a liquid-propane (LP) gas truck. Some trucks even have the capability of switching from LP gas to regular gasoline.

In the main, there is really very little difference in operating cost between the LP gas and the gasoline-operated truck. LP gas does extend engine life because it uses a cleaner-burning fuel; however, since LP requires additional components, such as converters, it does add cost to routine repairs. The additional costs largely offset any of the economic advantage gained by an extended engine life, and whether or not savings can be realized through the use of LP gas instead of gasoline might depend primarily upon fuel costs, which vary greatly from one location to another.

There is an advantage in using LP-gas vehicles inside warehouses and factories: their cleaner-burning fuel vents less noxious exhaust into the atmosphere. Regular gas-powered trucks are sometimes used on partially open receiving and shipping docks of warehouses and plants; however, their use is contingent upon exhaust systems to dissipate exhaust fumes, and even in these situations the desirability for improving working conditions is diminishing the number of gas trucks used in this manner.

Electric forklift trucks—i.e., those powered by large industrial batteries—have a higher initial cost (purchase cost), but have a much lower trade-in value, largely because only small companies are in the used truck market and most of these customers have only sporadic use for these trucks to perform the occasional heavy lift, such as loading or unloading freight. The added expense of batteries and charging equipment is not economically justifiable to them.

Electric trucks, although slower than either gas- or LP-propelled vehicles, offer the advantages of fume-free operation, which makes them ideal for indoor operation, such as in warehousing and manufacturing. In addition, electric forklift trucks are relatively more reliable than gas or LP vehicles, have lower maintenance costs, and are considerably less expensive to operate in these days of high fuel costs.

When you are considering gas, LP gas, or electric trucks, if it does not make any environmental difference—i.e., if either gas or electric trucks can be used in a particular operation—and you are only concerned with cost, then the following would apply:

A 1,000-running-hour level, which is common in many operations, would be 50 percent utilization of a truck in an 8-hour-day, five-day-per-week, one-shift operation. Many heavy-duty, high-running applications will keep a forklift truck occupied for 2,000, 2,500, or even 3,000 hours per year. Although the useful life of a forklift is generally considered to be about 10 years, a 3,000-hour-per-year equipment usage will shorten the economic life of a truck to 6 years for the electric and 5 years for the gas or LP truck. The electric truck's higher ownership cost will have been offset by lower fuel and maintenance costs and lower downtime costs.

Thus, it might be indicated from this discussion that electric-powered trucks should be used when high running hours are contemplated, and gas-powered equipment should be used for lower running-hour operations or intermittent usage. There are other factors involved, however, that could make electric-powered trucks the more economical choice, even for light-duty operations. Such factors might be the relative shortage of qualified forklift truck mechanics, high labor costs, or severe operating environments, such as a foundry.

Another factor to consider is the importance of downtime in an operation. If a truck must be sidelined because of malfunction, will it have a direct impact on the production of the plant? If so, then a more reliable truck, which requires less maintenance, might be the better alternative despite its higher ownership cost.

Labor requirements have an important bearing upon the total cost and the economic life of a forklift truck. There has been a crucial shortage of qualified forklift truck mechanics due in some part to the technological advances that have made the design of both gas- and electric-powered forklift trucks more complex. Many plant mechanics with automotive repair backgrounds are not fully qualified to handle the kind of forklift truck repair to keep the units in top operating condition.

All the major forklift truck manufacturers have attempted to ameliorate this situation by establishing factory training centers, seminars, and workshops to bring the plant maintenance mechanic up to a higher skill level. In addition, test and diagnostic equipment has been developed that will assist the mechanic in troubleshooting and in repair operations.

Another facet of the industrywide dilemma has been the higher degree of maintenance skill required for electric-forklift truck repair, in contrast to gas-powered equipment. Again the industry has attempted to resolve the difficulty through the use of electronic controls with plug-in components.

Although electric trucks have become more sophisticated, so have gas-powered units. In this regard the design of gas trucks has become more complicated with the introduction of high-performance, automatic transmission; powershifts; and electronic ignition systems. In view of the foregoing discussion, maximum forklift truck uptime at a minimum cost cannot be achieved using relatively unskilled maintenance personnel. As the skill level requirements rise, so also does the cost of labor.

All evidence points to the fact that maintenance requirements for gas-powered trucks are generally higher than for electric trucks. Computerized data files have added another dimension to this statistic. Maintenance cost composition shows a higher percentage of labor for gas-powered units over electric. On the average, maintenance costs for a gas truck are 60 percent labor and 40 percent parts; on the other hand, electric trucks with their more expensive components (that can be installed quickly) indicate labor costs that are approximately 50 percent labor and 50 percent parts. Therefore, as labor rates increase, electric forklift trucks, with lower maintenance requirements, should have an increasing cost advantage.

Of major significance to the materials management department is the effect that the operating environment might have on maintenance and downtime costs, and with that the question of whether gas or electric forklift trucks will prove more economical. Dusty, dirty, or corrosive environments have a severe effect on maintenance and downtime costs. If the cost of maintenance increases sharply per operating hour, the higher acquisition cost of electric forklift trucks is offset rapidly. The same would be true even if the cost of maintaining a gas-powered forklift truck and an electric truck both increased proportionately in severe environmental applications.

As a matter of fact, internal combustion engines such as gas and LP gas trucks are seriously affected by ingesting foreign particles, particulates, and dust of all kinds. Their carburetors suck in dirty air, causing throttles to stick. Also, dirty air is drawn into combustion chambers and blowby forces gritty abrasives into the lower end of the engine, causing heavy wear of journal bearings and piston rings.

Fortunately, electric forklift truck-drive motors can be effectively sealed against just such a dirty environment. It is a simple matter to seal the controls of an electric truck.

With a gas truck there are a number of relatively minor, but still damaging, problems that can be created by a dirty and dusty atmosphere—problems that are time-consuming and relatively expensive from a maintenance, labor-hour viewpoint. As an example, radiators become clogged and the operator doesn't even realize it until the engine overheats, sometimes seriously damaging the engine. Thus, in a severe environment, there are many minor problems, which when added together become cumulatively large, adversely affecting maintenance costs, downtime costs, the economic life of the vehicle, and the relative economics of gas-powered vs. electric-powered forklift trucks.

Nevertheless, fuel costs might very well be the single most important cost factor in the gas vs. electric forklift truck decision-making process, regardless of how the other factors align themselves. Fuel costs have skyrocketed as a result of the energy crisis the whole world is experiencing. Hardly a decade past, the difference in the cost of power between gas and electric was considered insignificant. Now, however, it is forming the very basis of a formidable industry shift in the type of equipment to be used.

In the final analysis, however, the following factors affect the gas vs. electric decision:

- Running hours
- Maintenance labor rates
- Downtime costs
- Operating environments
- Fuel costs

Simply because of the complexity of the many variables that control total costs, a computerized cost analysis might be the only way the individual plant can finally come to grips with the decision involved in the gas vs. electric forklift truck controversy. That is why the discussion in the previous section, on records and record-keeping, is of vital importance.

Equipment replacement justification   Whenever capital is to be invested—for example, in replacing MH equipment—there has to be an economic justification for doing so. Thus, over the course of years, the rule-of-thumb, or judgmental, method has given way to more reasonable, and ergo, more scientific methods. Therefore, there is at least more than one way to justify equipment replacement, all with a degree of validity.

An empirical method for justifying the replacement of equipment involves using the following:

- Job assignment
- Utilization
- Life cost per equipment hour
- Rotation of equipment to lesser jobs
- Installation of new equipment on the highest-use task
- Application of cost-per-hour figures at which the new unit should operate (estimated)
- Phasing out of a high-cost unit

Another way to justify equipment replacement is to use the increased-cost-per-hour method. If you plot the cost per hour in dollars on the vertical axis and the cost per thousand hours of operation on the horizontal axis, the coordinates of the point will establish a curve (which might require a little smoothing). By taking two successive points on the curve—for example, one point at 8,000 hours of operation and one at 9,000 hours of operation—we can say that the maintenance cost had increased from $0.50 to $0.75 an hour. Thus, owing to the high maintenance cost increase, the unit should be replaced. But the actual cost increase was not $0.25 per hour; it was more nearly like:

$$
\begin{array}{ll}
9{,}000 \text{ hrs} \times \$0.75 & = \$6{,}750 \\
8{,}000 \text{ hrs} \times \$0.50 & = \$4{,}000 \\
\hline
1{,}000 \text{ hrs} & = \$2{,}750
\end{array}
$$

or $2.75 per hour.

At 8,000 hours the unit was operating at $0.50 an hour, or $4,000 in maintenance costs. At 9,000 hours, the unit operated at $0.75 an hour, or $6,750 in maintenance costs. The difference, $2,750, divided by 1,000 hours is equivalent to $2.75 an hour.

Unfortunately in this example, replacement came after money had been expended on repair of the equipment.

A better method than the two methods just discussed, is called the lowest-total-cost point, or LTCP concept, in which each unit has its own individual cost curve. This total

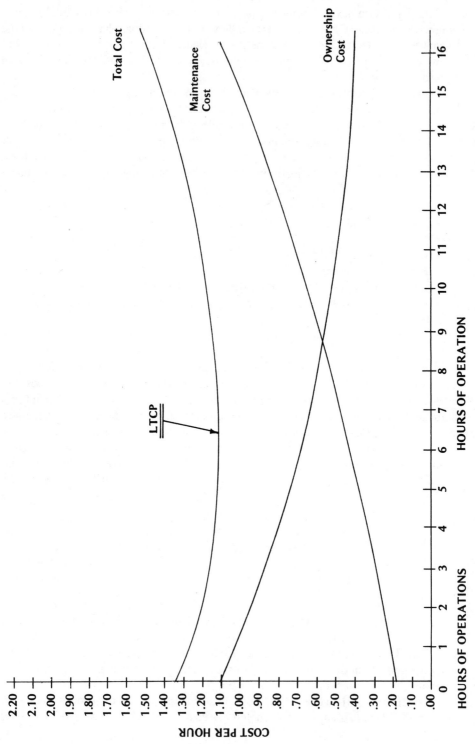

Fig. 12-10. The least total cost point (LTCP), which indicates when to replace a piece of materials-handling equipment.

cost curve is derived by adding the cumulative maintenance costs per hour and the cumulative ownership costs per hour and projecting (extrapolating) the total cost curve into the future. This extrapolated curve makes it possible to plan replacement prior to the need for major items of repair. (See FIG. 12-10.)

The ownership cost is a value derived by decreasing the acquisition cost at a quarterly declining rate (which does not include trade-in or residual value, taxes, or freight delivery charges). In this way, the ownership cost curve illustrates how much of the total cost of the unit has been used at the end of each quarter. The rate depends on a historical standardization process in which the type of unit and capacity for a particular manufacturer was developed after reviewing records of like equipment that had been replaced in order to obtain actual values.

The maintenance cost curve reflects the total amount of dollars spent for maintaining the unit in top operating condition. Also, in accumulating maintenance costs, all normal wear and tear items are included, but repairs caused solely by the operator's abuse are excluded.

The total cost curve for a particular unit is obtained by combining (adding) the maintenance cost curve with the ownership cost curve. What we obtain in this method is the point at which the unit will be operating at its lowest total cost. This is the point at which the unit should be replaced, for after this point has been reached, the total cost will begin to rise.

Looking only at ownership and maintenance cost, we have derived a simplified, but valid, approach to the replacement of equipment. Not included in this methodology is the cost of money, after-tax effect, operating cost, downtime, or obsolescence.

In an attempt to test the theory behind this concept, graphs were charted using known maintenance and ownership costs, applying the cost of money, the after-tax effect, and operating cost. With these new figures the total cost curve was changed to a point that gave an entirely unrealistic lowest total cost point, since the age of the equipment (unit) would indicate replacement more than the cost. Recalculating, using an empirical downtime cost factor again changed the total cost curve; however, it brought the LCTP back to its original position. After an examination of the simulated curves, it is felt that the original LCTP method is valid without any further changes. Also, since it is very difficult to assign a dollar value to obsolescence, this factor was not included in the calculations.

In pursuit of the subject of the effect of the cost of major overhauls on the curve, it has been found that if a unit has major work done to it, the quarter in which the cost was incurred will show a decided upswing of the total cost curve, whereas after repairs have been made, the unit should operate at a lower cost. Thus, the curve will reflect this lower cost. After four or five quarters, however, the total cost curve again will begin to swing upward and predict the LTCP. The upward movement of the total cost curve indicates that the maintenance cost curve is rising faster than the ownership cost curve is decreasing.

In a summation of the LTCP method, we have the following:

- Consistency   An approach for determining the point at which to trade in a piece of equipment. Instead of using a hit-or-miss method, the LTCP method gives fairly reliable guidelines.

- Eliminates costly overhauls and rebuilds   With LTCP charts, trends in equipment costs can be predicted so that costs can be extrapolated (forecast) making it possible to see how a unit will be performing for at least eight quarters—i.e., two years—into the future. If the unit appears to be approaching the LTCP, it is possible to avoid spending large sums of money for overhauls. If the method does not predict that the LTCP will be reached within the next eight quarters, then the decision to proceed with the overhaul is justified, rather than in waiting for a replacement unit.
- Minimize general maintenance   Since LTCP provides quarterly reports on maintenance and equipment utilization, it is possible to ferret out high-maintenance-cost units. This affords the opportunity of reviewing the specifications and operating conditions for the unit and to find out why the cost is high on a particular unit, or in a certain department of the plant.
- Assists in obtaining capital funding   By using the LTCP methodology, the materials management department obtains greater credibility for the justification of equipment replacement.
- Forecasts equipment requirements   It provides a tool to predict when certain pieces of equipment will need to be replaced. Therefore, a decision can be made to alert the equipment supplier to place it in his or her build schedule. By doing so, long delivery lead times and higher prices can be avoided. Also, it provides the necessary advance warning that capital dollars should be budgeted for this purpose.

Return on investment (R.O.I.) and the discounted cash flow method

In earlier chapters of this text, I have used the expression *return on investment* (R.O.I.). In general, the use of the term comes to the fore when we attempt to persuade the top management of the company to part with company funds for a project that we ardently believe to be an opportunity to improve the profitability of the enterprise—i.e., the company as a whole.

Thus, many companies have established capital expenditure limits for officials of their companies. For example, a divisional manager might have authority to approve expenditures of up to $2,500. His or her subordinates need only to prepare a one-page summary of benefits that will accrue to the company by means of this purchase. Expenditures of over $2,500 to possibly $25,000 might require a vice-president's approval with a more elaborate justification. However, when it comes to large investments involving hundreds of thousands, or millions, of dollars, it might require a staff of 10 to 15 working full time for a period of 6 to 18 months in order to provide the necessary justification for the expenditure, say a warehouse with an automatic storage and retrieval system with computerized control.

What management needs in order to approve this expenditure is, in brief:

- To establish the proper priority for the allocation of company dollars, which are always in short supply
- To obtain the money by borrowing or other means
- To obtain an R.O.I. that is capable of satisfying the stockholders of the company

- To make a management decision of the adequacy of the R.O.I. based upon the cost of the money to be invested, plus or minus a relatively small margin of error

Although there are many ways to determine the R.O.I., the following discussion concentrates on just two of these methods. By and large these two methods are the most common and accepted ways to indicate the potential return.

The first commonly used method can be expressed in this manner, using hypothetical figures:

$$Annual\ return = \frac{Yearly\ savings}{Investment} = \frac{\$3,000}{\$10,000} = 30\%$$

or a 30 percent per year return for this example.

Determining the R.O.I. by means of the Discounted Cash Flow Method is the second method commonly used. As indicated, there are usually many demands on a company for capital investment from many departments. It is logical to assume that it would choose the investment that will result in the largest rate of return.

Rate of return   Suppose there is a choice of two injection molding machines (or two types of lathes, etc.). Machine A can be purchased for $10,000 and will result in a savings of $10,000 per year for the next seven years. Machine B can be purchased for $20,000 and will save $15,000 per year for the next seven years.

As you can see Machine A will return 100 percent per year, and Machine B will return 75 percent per year, or:

$$Savings - cost = Return$$

Based on these returns, it looks like Machine A is the better choice; however, we might be too hasty. Let us examine Machine B again.

Machine B costs $20,000. If we take $10,000 (or the cost of Machine A) from Machine B and say that it will return 100 percent in the first year, we can say that the additional $10,000 will return $5,000, or 50 percent in the first year. If you cannot make an additional or alternate investment with the second $10,000 increment that will return the same amount, 50 percent, in the first year, then the second machine, B, would be the more logical investment of capital.

From this example, it is evident that there are a number of things to evaluate when considering the investment of capital. For instance, the rate of return might vary according to the choice of investment. In addition, however, there are other considerations.

Pay-back period   Another element to consider in capital investment is the *pay-back period*, which is the amount of time required for the return to equal the investment. As an example, let us say you are going to invest $100,000 in plant expansion and new equipment, and your projections for this investment indicate the annual return to be $10,000. Your company pays a tax rate of 48 percent, but you assume 50 percent to be conservative (and for easier calculation). Your pay-back period, before taxes are removed, equals 10 years ($100,000 ÷ $10,000). If you figure the pay-back period after taxes, then the pay-back period is 20 years (or $100,000 ÷ $5,000).

It is important to remember, when comparing different investment possibilities, that you must use the same ground rules for each investment; otherwise you are comparing apples to oranges and getting a biased result. For example, if the pay-back period is after taxes for one investment, it should be on this same basis for the other investment opportunities that are under consideration.

In another example, Investment X, consider buying a piece of equipment costing $10,000 with a life expectancy of ten years. The gross savings in the first year will be $38,000, but there will be no further savings after that year.

Let us perform the calculation, as follows:

Gross savings	$38,000
Less depreciation	2,000
Savings before taxes	36,000
Taxes @ 50%	18,000
Net savings	18,000
Add back depreciation	2,000
Cash Flow	$20,000

Pay-back period = 1 year

In this investment, we can invest $20,000 in a piece of equipment and obtain a $20,000 return in one year. Thereafter there are no further returns.

You will note that the depreciation was added back to the savings to arrive at a cash flow amount. Therefore, it is our cash flow that we are considering as our savings each year.

Since we are putting in $20,000 and taking out $20,000, we have not made any progress—we haven't gained anything. Therefore, we would consider this a poor investment, since we actually lost the use of $20,000 for one year, although it might be possible to sell the piece of equipment at the end of the year for $15,000 to $18,000 and receive some return.

Investment Y also demands the purchase of a $20,000 machine, again with a life of ten years. This alternate investment will produce a gross savings of $8,000 for several years. Let's examine it, as follows:

Gross savings	$8,000
Less depreciation	2,000
Savings before taxes	6,000
Taxes @ 50%	3,000
Net savings	3,000
Add back depreciation	2,000
Cash Flow	$5,000

Pay-back period = 4 years

In Investment Y our gross savings of $8,000 is decreased to a cash flow of only $5,000. It will take four years before we start making a profit on the capital we have invested.

Investment Z indicates an expenditure of $20,000, also, with a ten-year life for the equipment. This piece of equipment will produce the following savings:

1st year	= $ 2,000
2nd year	= $ 6,000
3rd year	= $10,000
Each additional year	= $16,000

In evaluating the pay-back period, you can see what will happen by reviewing TABLE 12-2. The $20,000 investment will be returned in four years.

Table 12-2. Pay-Back Period for Investment Z.

	1st Yr	2nd Yr	3rd Yr	Every Additional Year
Gross savings	$2,000	$6,000	$10,000	$16,000
Less 10% depreciation	2,000	2,000	2,000	2,000
Savings before taxes	0	4,000	8,000	14,000
Taxes @ 50%	0	2,000	4,000	7,000
Net savings	0	2,000	4,000	7,000
Add back depreciation	2,000	2,000	2,000	2,000
Cash flow	2,000	4,000	6,000	9,000
Pay-back period = 4 years				

Let's compare Investments X, Y, and Z. See TABLE 12-3.

Table 12-3. Comparison of Investments X, Y, and Z.

	Pay-Back Period	Cash Flow Each Additional Year
Investment X	1	$0
Investment Y	4	$5,000
Investment Z	4	$9,000

Each of these investments requires an amount of $20,000. So, let's calculate the percent of return on each investment after the pay-back period:

X = 0 or,

$$\frac{0}{20,000} = 0$$

Y = 25%   or,

$$\frac{5,000}{20,000} = 25\%$$

Z = 45%   or,

$$\frac{9,000}{20,000} = 45\%$$

It appears that Investment Z is the best choice.

Unfortunately, a difficulty with using only pay-back period calculations is that the length of the return is not factored into the equation. For example, if the 25 percent return for Investment Y was for the next five years, but the 45 percent return after pay-back of Investment Z was only for one more year, then we would have to admit that Investment Y is the best choice because the total return for Investment Y is $45,000, figured in this manner: Cash flow of $5,000/year for 9 years = $45,000. And the total return for Investment Z is $30,000, figured thus: cash flow in pay-back period = $21,000 plus $9,000 in the fifth year, $30,000.

The rate-of-return method of evaluating investments attempts to correct one of the problems of the pay-back period. It does so by taking into consideration the length of return on the investment.

For example, a company requires replacement of a crucial piece of equipment on its production line, and it has the choice of two alternate machines, each costing $20,000.

Machine A:

Cost	$20,000
Return/year	$5,000
Pay-back period	= 4 years

By using the rate-of-return method, we factor in the length of the return. Both machines have a life of ten years. The return per year of Machine A is $5,000. Therefore, the total return will be $50,000 ($5,000 × 10-year life).

$$\text{The total return in } \% = \frac{Total\ return}{Original\ investment}$$

or:

$$\frac{\$50,000}{\$20,000} = 250\%$$

Machine A will return 250 percent over its ten-year life. The average yearly rate then for Machine A is:

$$\frac{250\%}{10\ \text{years}} = 25\%\ \text{per year}$$

Machine B also costs $20,000, but its yearly return is as follows:

$$
\begin{aligned}
\text{1st year return} \quad &= \quad 0 \\
\text{2nd year return} \quad &= \$1,000 \\
\text{3rd year return} \quad &= \$3,000 \\
\text{Each additional year} \quad &= \$8,000
\end{aligned}
$$

The pay-back period for Machine B is five years.

If we add the sum of the returns for ten years of the Machine B's life, we obtain the following:

Year	Return
1	$ 0
2	1,000
3	3,000
4	8,000
5	8,000
6	8,000
7	8,000
8	8,000
9	8,000
10	8,000
Total Return	$60,000

Then,

$$
\frac{Total\ Return}{Original\ Investment} = \frac{\$60,000}{\$20,000} = 300\%
$$

and the average yearly return for Machine B is equal to:

$$
300\% \div 10 \text{ years} = 30\%
$$

We can summarize the investments for the two machines as shown in TABLE 12-4. It is now possible to make the decision to purchase Machine B.

Table 12-4. Comparison of Investments A and B.

	Machine A	Machine B
Investment	$20,000	$20,000
Pay-back period	4 years	5 years
Total return	250%	300%
Average return	25%	30%

It is obvious, therefore, that the advantage of the rate-of-return method rests in the fact that it takes into consideration the length of time over which a return can be expected. If we had used only the pay-back method, Machine A would have seemed the better investment of the two, since it showed a shorter pay-back period.

Present value concept   In the previous discussion, we assumed that the dollars used will remain at a fixed value from year to year. But, let us look at this in another way. Let us suppose that we have to choose between receiving $1,000 now or $1,000 one year from now. Wouldn't we choose the $1,000 now? Of course, we would. We could invest the $1,000 we receive now and have more than $1,000 a year from now.

Let us look at another example. Just suppose we are able to invest $10,000 in one of two investments, X and Y. The investments are shown in TABLE 12-5.

Table 12-5. Returns of x vs. y.

		1st Year	2nd Year	3rd Year
X	$10,000	$500	$300	$200
Y	$10,000	200	300	500

The total return in each of the two investments is the same; however, wouldn't we choose Investment X as the better one to make? Of course, we would! Investment X is the better choice because it has a larger first-year return of $500, which we can invest for a larger gain. This is a lot better than having only $200 to invest, as in the case of Investment Y.

It stands to reason that the money you receive in the future is not worth as much as the money you receive now. Therefore, in evaluating investment projects, you should discount, by a certain amount, money to be received in the future.

Let us consider investing one dollar for one year. At the end of the year that dollar placed in a bank at an interest rate of 5 percent would return to us $1.05. Or we could invest $0.952 now and, at 5 percent, draw out $1.00 one year from now. Therefore, the *present value* of the dollar we will receive one year from now is $.952, if discounted at the rate of 5 percent.

In order to minimize the amount of calculations that must be made, we find it very convenient to use TABLE 12-6. The left column indicates the number of periods during which interest is compounded on the investment.

From the table you can determine the value of one dollar for different investment periods and interest rates; for example:

$$6 \text{ years } @ \ 7\% \ = 0.6663$$
$$15 \text{ years } @ \ 8\% \ = 0.3152$$
$$20 \text{ years } @ \ 9\% \ = 0.1784$$
$$25 \text{ years } @ \ 10\% \ = 0.0923$$

Table 12-6. Present value of $1,00.

n	6%	7%	8%	9%	10%	11%
1	0.9434	0.9346	0.9259	0.9174	0.9091	0.9009
2	0.8900	0.8734	0.8573	0.8417	0.8264	0.8116
3	0.8396	0.8163	0.7938	0.7722	0.7513	0.7312
4	0.7921	0.7629	0.7350	0.7084	0.6830	0.6587
5	0.7473	0.7130	0.6806	0.6499	0.6209	0.5935
6	0.7050	0.6663	0.6302	0.5963	0.5645	0.5346
7	0.6651	0.6227	0.5835	0.5470	0.5132	0.4817
8	0.6274	0.5820	0.5403	0.5019	0.4665	0.4339
9	0.5919	0.5439	0.5002	0.4604	0.4241	0.3909
10	0.5584	0.5083	0.4632	0.4224	0.3855	0.3522
11	0.5268	0.4751	0.4289	0.3875	0.3505	0.3173
12	0.4970	0.4440	0.3971	0.3555	0.3186	0.2858
13	0.4688	0.4150	0.3677	0.3262	0.2897	0.2575
14	0.4423	0.3878	0.3405	0.2992	0.2633	0.2320
15	0.4173	0.3624	0.3521	0.2745	0.2394	0.2090
16	0.3936	0.3387	0.2919	0.2519	0.2176	0.1883
17	0.3714	0.3166	0.2703	0.2311	0.1978	0.1696
18	0.3503	0.3959	0.2502	0.2120	0.1799	0.1528
19	0.3305	0.2765	0.2317	0.1945	0.1635	0.1377
20	0.3118	0.2584	0.2145	0.1784	0.1486	0.1240
21	0.2942	0.2415	0.1987	0.1637	0.1351	0.1117
22	0.2775	0.2257	0.1839	0.1502	0.1228	0.1007
23	0.2618	0.2109	0.1703	0.1378	0.1117	0.0907
24	0.2470	0.1971	0.1577	0.1264	0.1015	0.0817
25	0.2330	0.1842	0.1460	0.1160	0.0923	0.0736
26	0.2198	0.1722	0.1352	0.1064	0.0839	0.0663
27	0.2074	0.1609	0.1252	0.0976	0.0763	0.0597
28	0.1956	0.1504	0.1159	0.0895	0.0693	0.0538
29	0.1846	0.1406	0.1073	0.0822	0.0630	0.0485
30	0.1741	0.1314	0.0994	0.0754	0.0573	0.0437
35	0.1301	0.0937	0.0676	0.0490	0.0356	0.0259
40	0.0972	0.0668	0.0460	0.0318	0.0221	0.0154
45	0.0727	0.0476	0.0313	0.0207	0.0137	0.0091
50	0.0543	0.0339	0.0213	0.0134	0.0085	0.0054

Let us use an example to apply this knowledge to two investment projects requiring the same initial amount of capital:

- Investment X returns $300 in two years.
- Investment Y returns $400 in four years.

Say, for example, that we discount both at 8 percent.

$$\text{Investment X} = 8\% \text{ for 2 years}$$
$$= 0.8573$$

$$\$300 \times 0.8573 = \$257.19$$

$$\text{Investment Y} = 8\% \text{ for 4 years}$$
$$= 0.7350$$

$$\$400 \times 0.7350 = \$294.00$$

Investment Y has the larger present value; therefore, it is the better of the two investments.

In another example, let us suppose that a company has to decide between two expansion projects. With Project A, it can invest $100,000 to expand its plant with certain high-production equipment. The company is very conservative and feels that this is a relatively safe investment that will return $30,000 the first year, $60,000 in the second year, and $60,000 in the third year. Because this is a relatively safe investment, the company decides to discount the investment at 6 percent, which is exactly the cost of acquiring the money to finance this expansion program.

The present value of the return on investment, if calculated, is, as follows:

$30,000 for one year @ 6% discount:	
$30,000 × 0.9434 =	$ 28,302
$60,000 for two years @ 6% discount:	
$50,000 × 0.8900 =	53,400
$60,000 for three years @ 6% discount:	
$60,000 × 0.8396 =	50,376
Present value =	$132,078

With Project B the company can install a new production line that will increase a certain product's share of the market. However, since it is a fairly new product and not so well established, there is a high degree of risk involved for the investment in launching this product.

The company, therefore, sets the discount at 30 percent, or five times the cost of obtaining the money. In other words, the risk is high, but the gain could be high also. Let's calculate the returns, as follows:

$40,000 for one year × 0.7692	=	$ 30,768
$80,000 for two years × 0.5917	=	47,336
$140,000 for three years × 0.4552	=	63,738
Present value	=	$141,832

Project A has a present value of $132,078, whereas Project B has a present value of $141,832.

Taking another example, let us suppose that a company will invest $20,351 in a project and receive the following returns:

- Year 1: $6,000
- Year 2: $8,000
- Year 3: $10,000

Let us calculate the yield, or rate of return, on this investment. In the first place, we need to find the rate of discount that causes the sum of the present values to equal the investment,. This is done on a trial-and-error basis; so, let's start by applying a discount rate of 6 percent to the returns and see what the sum of the present values turns out to be.

$$
\begin{aligned}
\$\ 6{,}000 \times 0.9434 &= \$\ 5{,}660 \\
\$\ 8{,}000 \times 0.8900 &= 7{,}120 \\
\$10{,}000 \times 0.8396 &= 8{,}396 \\
\text{Present value} &= \$21{,}176
\end{aligned}
$$

The present value of $21,176 is more than the original investment of $20,351; therefore, the discount wasn't large enough. Now, we'll try 10 percent and see what happens:

$$
\begin{aligned}
\$\ 6{,}000 \times 0.9091 &= \$\ 5{,}455 \\
\$\ 8{,}000 \times 0.8264 &= 6{,}611 \\
\$10{,}000 \times 0.7513 &= 7{,}513 \\
\text{Present value} &= \$19{,}579
\end{aligned}
$$

As you can see, the discounted rate of 10 percent gave us too low a figure, $19,579 vs. the $20,351 of our investment. So, we try 8 percent.

$$
\begin{aligned}
\$\ 6{,}000 \times 0.9259 &= \$\ 5{,}555 \\
\$\ 8{,}000 \times 0.8573 &= 6{,}858 \\
\$10{,}000 \times 0.7938 &= 7{,}938 \\
\text{Present value} &= \$20{,}351
\end{aligned}
$$

Our computations indicate, therefore, that the present values of the return at 8 percent are equivalent to our original investment. This investment bears the same return as if the investment had been deposited in a bank bearing an interest rate of 8 percent.

Notes

1. OSHA is the abbreviation for the Occupational Safety and Health Administration of the U.S. Department of Labor, which was established as a result of the Williams-Steiger Act of 1970 to safeguard the worker and the workplace.

13

Traffic and transportation

Introduction

In contrast to the purchasing department, which will buy anything that the various departments of the company require, the traffic department buys only one thing: transportation service. In the main, traffic is concerned with outbound shipments, although it also functions in specifying and arranging for in-bound shipment transportation. When the transportation function is added to the materials management department, which usually comprises purchasing, the added cost of transportation means that a large part of the company's funds will be administered by the materials manager. Thus, he or she will have the additional leverage in corporate matters that both traffic and purchasing bring into the boardroom, since the traffic function is a very important element in most industries.

In the typical manufacturing company, transportation services are the third greatest expenditure, being surpassed only by purchased materials and labor. Producers of low-cost, bulky materials often spend as much as 25 percent of their sales dollars on transportation services, and even companies that produce high-dollar-value items might expend as much as 5 to 10 percent of their sales volume on transportation.

For this reason, many of the larger companies have a separate traffic department, and the traffic manager is recognized as a major company executive. In some companies he or she may report directly to the president. However, in this text, I feel that as part of the activity responsible for material movement, the traffic function should be directly under the materials manager in the corporate structure, if we are to regard this function from a systems approach.

In the traffic department there are four major areas of responsibility:

- The selection of common carriers and the negotiation of rates for shipments. Despite the fact that a company might have a captive fleet of trucks, there is always the necessity for having to deal with common carriers.

- Tracing both in-bound and out-bound shipments using electronic data interchange (EDI) methods, or other means to track freight both for customers and in-plant departments such as purchasing and inventory control.
- Auditing freight charges and filing claims for refunds for excess charges or damaged shipments.
- Rotating carriers so that each carrier obtains a fair share of the company's business; developing new methods to reduce transportation costs; analyzing tariffs; negotiating special traffic arrangements, as required; and maintaining the security of shipments, maintaining truck and rail registers of incoming and outbound freight, and the like.

Many manufacturing and physical distribution companies operate their own fleet of trucks, not only to deliver end products and other merchandise to customers, but also to pick up purchased materials from their suppliers. In addition, companies that are not in the transportation business, per se, often operate their own ships, barges, aircraft, railroads, and pipelines. However, most companies rely almost entirely on common carriers for their transportation requirements.

In a business sense, the common carrier is like any other supplier that the company might have. The carrier performs a service for which the company must pay. The similarity ends there, however, because common carriers that cross over state boundaries are very restrictively legislated by the Interstate Commerce Commission (ICC), in addition to the usual parallelism that gives state agencies another method for bureaucratic control, regulation, and taxation.

This aspect of the traffic function gives it a unique character because of the complexity of the traffic rating structure. As an example of the rate regulation see: Uniform Freight Classification (UFC) "Ratings, Rules and Regulations"; National Motor Freight Classification (NMFC) "Classes and Rules" and Hazardous Materials Regulation of the Department of Transportation "Specifications for Shipping Containers."

For these reasons, special training and skills are required by traffic department personnel. Also, traffic is intimately involved with materials management, since the choice of a carrier often influences inventory stock levels. In this regard, speedy, reliable transportation service permits lower inventory levels to be maintained, and it is a crucial factor in just-in-time manufacturing. The traffic function is also related to warehousing and plant location, since the availability of low-cost transportation would be an important factor in plant location and warehousing in that its continuing requirement for price effectiveness depends, in part, on how well the traffic function is carried out, especially in (its) relation to carriers and contract haulers.

Receiving

Material flow through manufacturing or the physical distribution center begins when the material arrives at the plant's receiving platform. Prior to this time, the material might have had to have been packaged; traffic rates and routings, arranged; and the transportation mode, established. These latter activities are performed by the plant's traffic department. The rapidity with which the merchandise is removed from the carrier, counted, inspected, and logged in is a measure of the effectiveness of the receiving crew.

Receivables and truck/rail registers

Some of the physical aspects of the receiving function are: to identify incoming material, to verify this material, and to move this material expeditiously to the point of use, or storage, within the plant. In some companies, the material control handler will accept the merchandise from the assigned receiving department materials handler and transport it to its first point of use.

If the material is moved out of the receiving area, it is understood that it has been entered into (logged in) either the truck or rail register. Actually, this is one of the first nonphysical activities that is engaged in when the shipment arrives at the plant. In these registers, one of which is established for truck receivables, the following data is recorded:

- Truck time of arrival
- Date of arrival
- Driver's name
- Trucking company
- Bill of lading numbers
- Time unloading has been completed

The time when unloading has been completed is important when verifying demurrage charges. A very similar register, which is kept separately from the truck data, should be maintained for railroad car freight also.

The information from the truck and rail registers is important, also, from the standpoint of the traffic department, since it is this department's responsibility to deal with either the railroad or the trucking companies in the event that material is damaged in transit.

Weigh counting and other methods

It is in the best interest of the company to check materials on the receiving floor as rapidly as possible so that the carrier's bill of lading can be signed and the carrier released. Speed does not mean, however, that accuracy in counting incoming materials must be sacrificed. With automatic identification means, such as bar-coded labels, it is possible to establish counts for receivables quickly and accurately.

Therefore, the purchasing department will have included the requirement for bar-coded labels in each of its purchasing specifications. In the event that materials are received without bar-coded labels, the receiving clerk should have a prepared form that he or she can complete, which will indicate all of the pertinent data required for future purchases, and transmit a copy of this data to purchasing. The receiving clerk should note the following data:

- Part number
- Part name

- Date of receiving report
- Vendor name (and plant)

Not only should this data be transmitted to purchasing, but it should enter the receiving department's delinquent vendor file. On subsequent due-ins, the computer should be programmed to scan the delinquent file for this vendor's name. If the vendor is still not bar coding his or her shipments, then appropriate action should be taken. The purchasing department should have established a penalty in the vendor contract that would permit a price adjustment. If the shipment is a large one, then contract labor, etc., can be used to perform the bar coding of each vendor package.

Another method to speed-counting receivables involves small parts and scale weigh-counting methods. In this form of counting, a sample of the parts is counted and weighed. Multiples of this sample weight indicate the total count. With electronically operated weigh scales, the precision of the count is sufficient for payment purposes, since small parts usually do not have a very high unit cost.

By pursuing such approaches to counting, and by introducing load cells, conveyor photoelectric counting, and the like, it is possible to count large quantities of materials rapidly without incurring demurrage charges.

Shipping

The traffic department has the responsibility for buying transportation services, and it should buy it just like the company buys any other commodity, usually on quality, price, and service. As indicated previously, the traffic department should rotate carriers in order to distribute the company's transportation business in a manner that is fair to all carriers. Nevertheless, a company is not only interested in the price of the service, but more often than not it is more interested in the quality of service.

Usually in this type of business, *quality* means response time and delivery time. In other words did the trucker come as soon as called? Was the freight car available when requested? Usually, there is more urgency involved with motor freight, and the over-the-road trucker is much more flexible than the railroads, so much of this discussion pertains to the motor freight carriers. Carriers that give the fastest service sometimes charge higher rates than do the companies providing the slowest service; however, it is fortunate that this is not always true.

One of the problems involved with railroad car shipments is that larger quantities must be shipped longer distances to make rail shipment really pay for itself. Therefore, in the case of relatively large quantities of freight, the time element is rarely as urgent as it would be with the faster and more expensive motor freight haulage.

Another problem with rail shipments is that the customer must either have a railroad siding at the plant, or a piggyback shipment must be arranged. A *piggyback* is a motor truckload that is transported by rail to take advantage of the lower, long-haul freight rates. In lieu of this transportation mode alone, provision can be made to have a local motor carrier pick up the load at the plant and deliver it to the train's team tracks, or the customer can make the delivery to the train.

The traffic department has many options for shipment modes. Two of these have just been discussed. However, in addition to motor freight and rail, there are: freight forwarders who will pick up in their trucks and make deliveries by truck, air, ship, and barge; the Railway Express company; air freight; air express; air freight forwarder; U.S. Parcel Post; and United Parcel Service.

In addition to the many transportation modes available to the traffic department, a number of other characteristics to the shipping task also affect the price of the service, one of which is the routing of the carrier. There are almost always at least two possibilities for getting from point A to point B,. and more than likely, there might be dozens of ways to proceed from Seattle, WA, to Greensboro, N.C., Chicago, St. Louis, and Denver. These cities might be terminal points for quite a few carriers; thus, the selection of possible routings requires skill and experience on the part of the traffic technician.

Large shippers often find it convenient to reroute full cars or truckloads in transit. As an example, a large manufacturer might have a couple of hundred railcars en route from one of its major manufacturing plants in the midwest to its assembly plants in Tennessee, California, Oregon, and Texas. If shortages should happen to develop at any one of its assembly plants, then a car could be rerouted in transit to that particular plant. This traffic management function cuts lead times substantially and permits the company to save money because of lower inventory investments.

Another factor entering into the shipping world today that has many traffic managers worried has to do with city deliveries. Where a manufacturer or physical distribution center must deliver to a large number of drop points in large metropolitan areas, the vehicular glut is becoming especially severe, and sometimes during rush hours there appears a new phenomenon called *gridlock*, where traffic crawls to a standstill. The problem is becoming so severe that large numbers of city delivery carriers can make only one round-trip daily, and sometimes must even incur overtime to complete a run.

One large manufacturer in the automotive field has prevailed on its dealer-distributors to accept deliveries at night; that is, after normal business hours. There was some reluctance at first to accept this concept, but when provision was made to have the drivers place the delivered shipment in an enclosed locked area at each distributorship, the reluctance turned to a general acceptance of the idea. As a result of this innovation in shipping methodology, the manufacturer's fuel costs dropped, and overtime was offset by only a small amount of premium pay for the graveyard shift. The distributors' customers liked the concept because it meant they didn't have to wait an additional day for service.

14

Internal and external distribution networks

Internal networks

When finished end products are moved from production and assembly lines in a manufacturing plant to a company warehouse, or from the end of the company's production lines through distribution channels to the ultimate consumer, the process is called *physical distribution*. The physical distribution manager might also be responsible for the physical movement of partially finished products, such as spare parts, within the plant and might receive and transport purchased materials to and from storage areas and to their point of use, also within the plant. Eventually, however, the physical distribution process involves getting the product out of the plant or distribution center to the user.

Sometimes the traffic function is included in the physical distribution organization, but more than likely, as is suggested by this text, it is a part of the materials management system. In some companies, where the transportation and storage of merchandise, especially finished goods, is not particularly costly, the physical distribution department might be an independent function. This is especially true in nonmanufacturing companies, where the integration of a number of complex departmental functions is not an important criterion for observing a systems approach.

The same philosophy of regarding physical distribution as a unique entity has a parallel in the accounting aspect of functions that concern direct and indirect labor. For example, in a physical distribution center that is not a part of a manufacturing plant, the chart of accounts would most likely label direct labor as the functions concerned with four main areas of the facility: receiving; shipping; order selection; and packing. Thus, when physical distribution is a part of a manufacturing enterprise, all of these functions would be shifted into the category of indirect labor, as would many, if not all, of the material control functions that exist merely to support the manufacturing operations that would be categorized as direct labor if they were involved directly with the production of a piece part. The machine operator becomes direct labor in a manufacturing plant, and all of the support functions would be termed indirect.

In internal distribution operations, as in a typical warehouse, there are three kinds of order selection (picking) operations:

- *Bulk picking* Where the load of goods or end products are placed into storage and selected from storage in bulk quantities—i.e., in the same container, skid, pallet, or unitized load.
- *Break-bulk* When unit loads are broken down, and portions of the load are selected out of, or from, the larger unit load.
- *Bin picking* In which bulk merchandise from either bulk storage or break-bulk storage is broken down into smaller, individual cartons or pieces, and usually hand-, bar-code-, or weigh-counted (as in the case of fasteners and other very small parts that are not packaged individually).

In all three kinds of order-picking activities, the amount of inventory control that is practiced is usually indicative of the effectiveness of the operation. As with in-process control in the manufacturing area, the key word in this area is *control*, and it is suggested that a basic ingredient in physical distribution activities be the measure of control that can be applied in order that the operation as a whole is profitable.

In addition to the control of inventory in the distribution process, the type of materials-handling equipment used in order-selection activities has a large effect in determining the productivity of this operation. The materials management department should exercise the same care and diligence to employ and equip its employees in this area with the best means (equipment) to perform their tasks commensurate with the satisfactory rate of return on the capital invested—i.e., used to acquire the equipment.

Equipment can be either purchased or leased, or a combination of lease-purchase can be negotiated with most suppliers of equipment, be it powered or otherwise. In some circumstances there are advantages inherent in leasing equipment, but you are sure to require the advice of a qualified accountant in arriving at this decision, since corporate tax practices might largely influence any such decision.

It is a well-known fact that a combination of methods and equipment that gives excellent results in one locale might not perform as well or produce equally satisfactory results in another area. The quality and the general environmental background of employees, including supervisory staff, are often the factors most likely to influence productivity levels.

Although there are regional variations in the manner in which equipment is used or abused and there are often differences in attitude and work ethic, one of the cardinal rules of materials management (management, per se) is the necessity of keeping employees informed regarding all contemplated changes in methods and equipment. It is astonishing that, regardless of company size, the inadvertent laxity, or lack of probity, that causes one management level to issue orders and instructions to a lower management level, or to the employees in general, without the preparatory spade work of preparing the climate for change has a very unsettling effect on the labor force as a whole. It is especially important where the facilities are unionized and the organized labor group has the capability of shutting down the facility, primarily because the mechanism for mass action is so well oiled.

That is one of the main reasons that, should a facility have a union, shop stewards and higher union officials of the labor group should be involved in the early phases of planned changes. A goodly amount of so-called labor unrest is caused unnecessarily by the unwitting gap in communications that might touch off a fire in labor relations.

It would be naive to expect that all lack of communications is caused by an insensitivity to union hubris. It is a well-known stratagem in some corporate areas to foment this type of conflict when the real intention of the company's management is to close a plant, to ride out a slump in the economy, or to use some such ploy as a hedge when a major supplier has been struck and production materials are unavailable.

Barring these dishonest activities, it might be that a lack of common courtesy has occasionally pervaded some segments of industry so that a largely antagonistic labor-management relationship has come about. This is especially evident in some segments of the automotive and aircraft manufacturing industries. Even in some of the health care and educational fields, where you would think that this medieval attitude would long since have disappeared, there is an overtly expressed antagonism that is hard to explain in the light of the many advances civilization has made in science and technology.

Since management depends upon labor for its productive efforts and labor requires the most effective implementation of managerial expertise for its combined success, it would seem entirely logical that, in order for both factions to carry out their tasks at the highest possible level of profitability, the communication gap be filled in. There is no better place to commence this effort than management's adherence to a philosophy of supplying information whenever there is a "need to know."

External networks

The materials management department has an obligation to assist the physical distribution segment of its network with all tools and materials it requires to perform its several tasks satisfactorily. In the previous section, the emphasis was on equipment to be used in the in-plant situation. There are several areas in the external facet of distribution that also require attention: the methods and materials used for packaging the product, and the preparation and means required to ensure the safe arrival of the product at the customer's place of business.

If the company has its own captive fleet of vehicles, then it can achieve savings in the area of packaging, since it could conceivably use corrugated packaging cartons of 150-pound test corrugated, instead of, say, 200-pound test, which would be required if it were to ship by common carrier. The reason for the heavier board is that the common carrier is liable for in-transit damage, and they insist that the goods be securely packaged. The 200-pound corrugated board is better able to sustain the rigors of transport. Naturally, the size, fragility, and weight of the object to be shipped are determining factors in packaging.

Another reason for using company-owned trucks is that the in-transit damage and the concealed damage is usually much less than when common carriers are the means for transportation. This kind of damage is especially prevalent when less-than-truckload (LTL) shipments are moved. It is not unusual for trucking companies to pick up LTL shipments at the company's shipping dock and transport the goods to a local terminal, where it is consolidated with other loads to make up a full truckload. The company's LTL shipment can be unloaded from the pick-up vehicle and placed on the trucking company's terminal floor, where it may wait a day or more for consolidation. In this interim it might be moved several times on the floor for physical rearrangement with other shipments. Finally, it will be loaded into an over-the-road vehicle, usually a tractor-trailer, and transported to the next freight terminal on the routing.

If the shipper is lucky, the material will remain on this truck; however, this does not always happen, and the goods might be unloaded at the next terminal to await further consolidation. Sometimes merchandise consigned to one carrier will be unloaded and reloaded into another carrier's vehicle simply because interstate regulations will not permit a tractor-trailer unlicensed in that state to operate in its jurisdiction.

As you can see, there might be several loadings and unloadings of merchandise in LTL-type shipments. The packaging of the materials must be capable of withstanding the rigors of fairly rough treatment, in addition to the shock and impact on the load primarily due to the truck that carries the material.

Thus, it is reason enough, in light of transportation modes that are available to a shipper, to ensure that the packaging of the product is accomplished as effectively as possible. The materials management department should have on its staff, whenever budgets permit, a packaging engineering department, or even one packaging technician—depending on the volume of shipments. A good packaging engineer can save the company many times his or her salary annually in savings from effective packaging methods.

Another area that packaging and materials handling engineers can effectively combine their talents and expertise is in the area of the design of containers—special shipping racks, and the like—that are customized to the company's products. The company can equip its company-owned vehicles with special handling devices that will not only make loading and unloading of the vehicles faster, but will protect the merchandise from in-transit damage. These are all measures that make the systems approach to materials management work well.

Company-owned vehicles are not the only transportation means in which innovative shipping strategies can be carried out. It is often possible to have contract haulers equip their vehicles, often at their own expense, in order to accomplish the same goals—the saving of the company's transportation dollars, the in-transit protection of product, and the enhancement or improved productivity of loading and unloading operations.

Quality objectives

As part of materials management's function, the administration of its various divisions requires that levels of performance be established that are consistently high, but attainable. In establishing these standards of performance, it makes good psychological sense if they are created by mutual consent—i.e., the person or group that has the responsibility for performing according to the standard should have some say in deciding what the standard should be like. Thus, it is important that job standards be on a par with job responsibilities. Like job descriptions, in a well-organized company the standards should be clearly written, and provisions should be made to make them a part of the job description.

When you consider the various aspects of physical distribution, then, from a systems viewpoint, there is a fine line between job description, employee performance, and the condition in which the materials, prepared by the distribution group, arrive at the customer's place of business.

Since quality is not a one-time thing, but must be sought after and carefully nurtured throughout the entire chain, there are many ways and places your merchandise might be damaged from the time it leaves your plant to the time when it arrives at the customer's door. Since you are concerned about your product and the good reputation of your company, there are several ways you have to ensure the safe arrival of the product. One of the ways is to use shock, impact recorders concealed in the packaging, which will alert the customer, and through the customer the company, of the presence of rough handling in the transportation mode, especially when items of high dollar value are shipped.

Another method, which is usually reserved for full truck and carload shipments, is the use of strain gauges on packages. This method also usually pertains to high-dollar-value shipments because of the instrumentation that is necessary and the technical labor that is required to set up the test. Also, when strain gauges are used, because of their overt nature, protruding wires, etc., it is a method that can only be used most satisfactorily when both the shipper and the carrier form a task force to perform the experiment. The use of these devices, then, has another value, which is largely psychological, since the exercise of such care and diligence often impresses upon both the company's and the carrier's personnel the necessity for handling the merchandise with care. In the final analysis, that is the goal worth striving for.

15

The use of facilities

Manufacturing processing

One of the key elements in obtaining the least total cost of materials handling in manufacturing plants is layout. Thus, one of materials management's concerns is to be able to influence plant layout to the extent that the effectiveness of materials control is optimized. There are other reasons, of course, to increase the efficiency of layout planning, in addition to that of increasing the productivity of the materials handling work force, some of which are the high cost of factory and warehouse space, including land costs and labor. Beyond these factors, there is also the dislocation and work interruption that are caused by the necessity to change a layout or create new space by plant expansion if the original layout was defective.

Despite these compelling reasons to do a good job of layout planning at the time of the first effort, it is sometimes necessary to reexamine the plant layout to make certain that some of the constant rigging and rerigging of production lines hasn't completely destroyed what had originally been conceived as straight-line handling.

One of the factors that makes layout planning so important is that if the shortest and most direct materials-handling route is not arrived at during the planning process, the circuitous path will have to be repeated over and over during the course of the year so that thousands of repetitions are made and many unnecessary hours are added to the payroll that could have been avoided. It is the cost-avoidance process that demands a review of layout and handling methods periodically in order to weed out the discrepancies in layout that usually manage to infiltrate even the best-laid plans. Therefore, if the shortest and most direct manner of handling is not practical, each increment of time wasted by handlers and equipment builds up rapidly, subtracting from the total effectiveness and profitability of the business.

Since materials-handling labor in the manufacturing end of the enterprise is labeled as "indirect," it is only necessary to examine plant records to note that there have been large increases in indirect labor-hours and cost in most plants. But accelerating indirect labor cost is not the only disadvantage in a shoddy layout. There is also a chain reaction in increased wear and tear on MH equipment and an increase in maintenance costs.

Outside of the obvious fact that a layout is required whenever a new plant or plant expansion is in progress, layout planning should be regarded as a continuing process for some of the reasons just outlined. Thus, the materials management department must be continually on the alert for any contemplated changes to the plant that are in the planning stages, in order that the materials-handling experts in the department can have a hand in the formation of reasonable and logical materials-handling routing and method.

It seems practical that there should be a collection of data prior to the planning stage of such things as volumes, capacities, kinds of equipment to be used, parts and products to be produced, production and fabrication methods to be used, and so forth, as the skills and experience of process planning and MH specialists dictate. This sequence of events should be followed—i.e., first data collection, then planning and layout for either new plant or plant expansion.

The layout planner might consider himself or herself fortunate if the operations closest to the proposed expansion are compatible. It sometimes happens that the planners are hampered by having operations of such large capital investment, such as a heat-treating furnace, soaking pits, or a paint line on the only side of the building that contains suitable space for the intended expansion.

One of the reasons that materials management should have an active part in all new plant planning is so that they can envisage future requirements, especially regarding materials-control functions. Site planning that allows for the proper and orderly expansion of the future plant might take longer and cost a little bit more, but it will prove relatively inexpensive in the long run.

Although it appears unfair to the top management of some companies to suggest that they sometimes vacillate in giving the decision to start building, it often happens that the company's design and construction (D&C) group will prepare drawings based upon an estimate of the required area—a preliminary estimate obtained from the planning group that represented a consensus.

Let us say for the sake of argument that this estimate is 500,000 square feet. Using this arbitrary figure, the D&C group will prepare drawings, and as soon as the drawings have been completed they are sent out for bid purposes. There is a valid reason for doing so quickly. Since construction prices tend to rise every year, the D&C people feel that to delay "going out on the street" will increase the cost of the building. Unfortunately, when D&C prepares drawings based upon 500,000 square feet, they have shaped the building into a rectangular structure 500 feet wide by 1000 feet long with a bay size of 50 feet square—i.e., a 50-foot distance on column centers.

In the interim, the layout planners are using machine templates and templates of handling equipment, cranes, conveyors, and the like, trying to optimize the flow of materials through the facility. They might even be using computer graphics and simulation. In the course of this planning several workable layouts are prepared.

Then the D&C group and the planners have a project coordination meeting, and the planners indicate that their best layout occurs in a space 400 feet wide by 1200 feet long. During the stunned silence that ensues, the D&C project manager unfolds roll after roll of drawings: site plans, sanitary lines, foundation plans, HVAC (heating, ventilation, and air-conditioning) drawings, structural steel drawings, water lines, gas lines, lighting, and generator stations. The dimensions are cast in concrete and frozen into steel weldments.

There are any number of layout planners, D&C groups, and architects and engineers in the larger companies who have experienced some of these nightmarish situations. Large companies, especially those in the heavy-industry segment, have so many special requirements for pits, infloor conveyors, heavy craneways, machine foundations that might be twenty feet or more below floor level, and the like that the D&C effort is very closely coordinated with the planning group so that when D&C finally sends the drawings of the plant out for bid, it usually represents a mutually derived schema of the manufacturing plant, regardless of how much time it takes to complete the drawings.

In all of the planning and processing that takes place in today's industry, one of the salient features should be the capability of looking into the future and anticipating the need for plant expansion whenever a major layout is prepared. It becomes very costly to replace toilets, locker rooms with tiled shower walls and floors, sewer lines, standpipes, and the like. More costly to relocate are power substations, railroad lines, water towers, fuel tanks, cooling ponds, waste-water reclamation equipment, and service roads—to mention only a few of the possible problems that can vex the future planner.

There are many opportunities to use foresight in planning a facility, just as there might be numerous restrictions confronting the planner. The type and configuration of the terrain might, in large part, dictate the size and shape of the proposed structure, and whether it is to be located in a suburban setting or in some sparsely developed country area. Neighboring residential communities, the public image the company desires to project, and the restrictions often imposed by local planning and zoning commissions have an immense impact on the kind of facility that might finally evolve. The process of site selection is almost an arcane science since so many factors are involved, not the least of which are water supply, sewage disposal, soil characteristics for footings and foundations, labor force, roads, available transportation, community activities, schools, and other amenities.

In some urban communities, it might be necessary to build a multistoried building if land values are very high. Also, in some electronic and appliance segments of the industry, multilevel buildings offer the materials management department a materials-handling bonus where component manufacturing is concerned. A good example is the Appliance Park built from the ground up in Columbia, Maryland, on the perimeter of the city of Baltimore. Smaller companies are sometimes forced into the older, multistoried loft buildings because of the cost of locating into new, suburban facilities. In this instance price alone dictates the type of facility that the materials manager inherits.

It is a real challenge to materials-handling ingenuity to fully utilize these older buildings, where ceiling heights are low—usually only 10 to 12 feet high in the upper stories—columnar spacing might be 20 to 25 feet on centers, and freight elevator capacities are fairly low.

It is buildings like these that tax the cost accountant's credibilities where the relatively inexpensive rental (or purchase) costs must be weighed against expensive elevator modification, downtime and maintenance costs, wasted space, and higher labor costs. Labor costs are often increased in loft building operations because a lot of the merchandise must be handled manually, handstacked because of the low ceiling heights and the inability to stack materials more than one or two pallet loads high.

Again, low elevator capacities might militate against transporting heavy riding forklift trucks to upper stories of the building. Floor loadings also might be limiting factors in the type of manufacturing or warehousing that can be conducted in loft-type buildings.

Aside from the kinds of buildings that can be used to house a company's processing or distribution activities, the question of site selection is most often considered the sole bailiwick of architects, land surveyors, architect-engineers (A/Es), land developers, and the like. Nevertheless, although the preparation of contour maps, delineation of drainage areas, and so forth require specialized technical skills, the materials management department, and the materials-handling practitioners in particular, should be as involved in the site-selection process as any A/E staff member.

Sometimes architects are more considerate of the aesthetics of the site than they are of the practical, more mundane aspects of materials handling. As an example, consider the railroad spur that cuts across an open field to curve alongside the plant building, hugging one whole side. It might be a potential problem area, after a few years, when the prospect of plant expansion is being considered and the expense of relocating the railroad spur limits the direction of plant growth.

Again, if the building is large, fire underwriters might insist that fire walls be placed in such a way that the building is compartmentalized into smaller segments so that fires can be contained. If the placement of these walls is done by the A/E they might be symmetrically disposed throughout important processing areas. Thus it is in such situations that the materials-handling engineer might more than earn his or her salary in recommending alternate divisions of the interior space of the building and substituting fire curtains, ceiling monitors, and sprinklers.

Site planning in relatively crowded urban areas offers further challenges to both A/Es and materials-handling engineers. Traffic and the size of transportation equipment make maneuvering into and out of plants using urban streets relatively difficult. Off-street parking is sometimes required, and fire ordinances offer further obstacles that must be overcome by the project engineer and planners. Inside the urban-located building, a number of other restrictive ordinances sometimes confront the layout planner, one of which is that a three-foot-wide aisle be maintained between storage areas and exterior walls.

These are only a few of the many factors that must be considered in the planning stages of manufacturing processing, and a few of the salient features are outlined as follows:

- Available water
- Sewage disposal
- Soil characteristics to permit footings without special construction
- Access for rail and trucks
- Potential for expansion
- Column spacing

- Fire safety
- Community demographics
- Labor force

This list can go on and on. That is one of the main reasons for materials management involvement, since the size and layout arrangement of the manufacturing plant (and warehouse or distribution center) cannot be determined by projected sales volume alone. The functions within the building must be considered, for they will directly affect the amount of space required for processing. The external and internal features of the structure are interrelated and demand equal consideration because the factory (building) must fit the requirements of the manufacturing operations.

In new buildings, two of the important considerations are bay size and column spacing. Columns should be located so as not to interfere with the movement and storage of materials. Columnar spacing, in either direction, that is too close will tend to restrict space utilization and reduce the effectiveness of materials-handling equipment. It is all too true that the type of construction might determine the spacing of columns, and lower-cost construction with limited free spans might result in higher operating costs. The best advice in this regard is to get the widest free span that money can buy and try to contain columns within building walls wherever possible.

Another caveat has to do with materials-handling equipment. During the planning stages of the facility, along with processing, the type of materials-handling equipment that is to be used should be considered carefully. If forklifts and pallets are the major equipment types, then doors must be made sufficiently high to accommodate the mast heights of the vehicles, and the load widths should dictate the door widths to be specified. Adequate aisle space must be planned for each processing area and workstation arrangement.

Ceiling heights and superstructure load-bearing strength should be adequate to accommodate conveyor systems, and column structures should be heavy enough for crane runways, if they are to be part of the handling system. Ceiling clearance should be provided so that the necessary 18 inches below the sprinkler heads are usable space. Fire underwriters require that no stacking of loads be permitted within this 18-inch limit.

Even the size of pallets and the containers to be used should be predetermined in the space layouts. The number of pallets and containers (or sizes of conveyors, when used) at each workstation should be part of the materials management contribution to the layout plan.

The group of individuals who has the task of laying out the new plant and doing the processing has a great deal of responsibility thrust upon it. Members of this group must, of course, be capable of working together because each worker brings his or her own expertise to the group. One of the most salient characteristics of these people is their ability to accept compromise. Often, during the course of the task group's working day, it will be necessary to weigh advantage against disadvantage; therefore, each team member must become familiar with all the operations to be performed in the new facility so that he or she will have a firm basis for assisting in the decision-making process.

In a new plant, it is well to question all the past processing methods, equipment, and so forth, of the earlier, predecessor facilities, simply because the goal is to avoid making the same mistakes of the precursor attempts at enhancing productivity. Almost everything about the new layout must be questioned so that what evolves is the best compilation of concepts and methodologies for the new facility. The planners have to keep in mind the capacity that the plant is being designed for and the amount of growth and volume that might be possible. It should be possible, although not easy, to envisage the plant at successive intervals of several years—five years and perhaps ten years into the future.

The basis for some of this advance planning would come from marketing and some from the combined effort of the task group to foresee technological development in their fields. The parameters that will supply information will be composed of the present processing methods and state-of-the-art technology, and the future state in which the plant is expected to operate.

The obvious sources of information that can be used to obtain the necessary background information are trade journals, trade associations, equipment manufacturers, and competitors' plants. Since the members of the task force for the new facility are the most alert "live wires" in the company, it stands to reason that all of them will be members of at least one industry trade association. Each of them will subscribe to at least one, if not more, of the vast number of trade journals that are available. The equipment manufacturers will be most happy to provide you, as a task group member, with all of the information on new processes and equipment that is available.

As far as knowing what is going on in competitors' plants, even the trade journals are good sources for this type of information. It is not expected that any task group member should be privy to any of a competitor's proprietary information on methods or processes because that would be tainted with the aura of industrial espionage. The materials-handling achievements of competitors, however, are often written up in the trade periodicals, and so gives increased credibility for any new methods that might be considered by the members of the materials management task force.

Engineering and testing

In the main, test cells are used primarily in heavy industry, where components are stressed and put through the environmental and operating rigors that they will encounter as part of the end product, if the item being tested is not the end product itself. If the test cells are to be located conveniently, then some portion of the new facility should be set aside from the main current of movement in the plant, but sufficiently close that the utilities of the plant should be capable of supporting this function adequately. Sometimes, however, test cells, because of the combustion effects, such as firing up jet engines in dynamic tests, must be located in somewhat isolated areas. Testing of internal combustion engines, for example, can be safely accomplished within the plant structure itself.

Since testing is one of the functions of the engineering department, the facilities to accommodate this activity should be located conveniently to the test engineering group of the new complex. A small conference room, which has been sound-proofed, should be provided for this activity, and the amenities should reflect the nature of the research and development effort—i.e., computer terminals, plotting boards, and other devices necessary for the engineering staff.

Another important consideration is the materials-handling methodology to be employed in obtaining test units from the production and assembly lines without interrupting the work flow. Also, the handling methods for getting test items into and out of the test cells, either manually or automatically, must be resolved in conjunction with the manufacturing and engineering departments. Sometimes it is even possible to perform all of the electrical, hydraulic, and mechanical hookups automatically without very much, if any, human interface. The scale of the test cell, naturally, should be commensurate with the size of the enterprise and the product.

Quality control

In any facility, whether it be an older plant or a project in the planning stage, it is important to locate the service facilities, inspection, and quality control (QC) departments in areas that are readily accessible to the functions they serve. In the main, the inspection and QC laboratory should be reasonably close to the receiving department floor or the receiving setout area to facilitate the taking of samples, the testing and analysis of specimens, and the like. In both older plants and new facilities, the installation of a laboratory is so expensive that it has to be permanently fixed in its relative location in the layout.

Aside from the receiving functional responsibilities, which would be all that would be required in a nonmanufacturing environment, the manufacturing responsibilities are usually very much improvisations. Some plants even maintain roaming QC inspectors, who wander from workstation to workstation on the factory floor.

In well-structured organizations, however, each production and assembly line might have to provide a QC workstation, where parts are inspected as they pass by, using either 100 percent sampling methods, where each and every part is inspected, or random sampling methods, where only a small percentage of parts is actually gauged and examined.

One of the new-facility task group's decisions might very well concern the use of fully automatic or semi-automatic testing diagnostics, in lieu of the slower, manual, and labor-intensive methods. Today's machine tools have become so sophisticated and complex that many not only have built-in diagnostics and change their own tools, but perform many of the testing and inspection functions as part of the processing sequence. Since every added machine tool feature increases the cost of the machine, a trade-off between the cost of inspection labor vs. automatic inspection becomes a consideration to be analyzed by the task group.

Maintenance

In a new facility, the plant maintenance (machine, not custodial) area should be located as closely to the forklift truck center of population as it is possible to get. This is true also of the forklift truck-battery-charging area if electric trucks are used. In the relatively closed, environmental atmosphere of a manufacturing plant, it is probably more desirable to use electric, rather than gas trucks. Even LP gas trucks leave plumes of fumes in their wake, and employee dissatisfactions increase when any type of gas mobile vehicles is used in close proximity to workstations.

In any plan for an electric-battery charging area, consideration should be given to use short roller conveyors to facilitate getting batteries in and out of forklift trucks. In lieu of these conveyors, jib cranes can be used, although hoists with chain falls are a considerable hazard in handling electric batteries because of the danger of shorting the cells.

Since battery-charging requires so much electrical energy, special wiring and conduit are necessary to install them. Thus, once the location of the battery-charging area with its special wiring and sewer connections has been selected and the equipment installed, it is, in some respects, similar to the QC lab in that relocation can be very expensive.

16

Information processing
and data entry

Decision-support systems

Since the costs associated with computers and peripheral equipment have continued to spiral downward, the many uses that the computer has in manufacturing and nonmanufacturing industries has trended upward. It is for no small reason, then, that computers are used in planning, management, and the control of manufacturing and distribution, as well as in other areas of nonmanufacturing operations.

Most of the following discussion refers to the term *manufacturing operations*; however, you should simply fill in for that term any of the nonmanufacturing substitutions you wish, since the arguments contained herein apply almost equally to these other avenues of commerce and industry. So when it is written that there are roughly three tiers of manufacturing where the computer is found, it would mean, just as likely, that these places might also be in nonmanufacturing segments of our economy.

Thus, at the top are the general and administrative functions of the company. It would depend on the size of the company, of course, as to whether or not a mainframe or large-scale computer would handle the planning activities and a wide panoply of accounting responsibilities; however, all the top executives and middle-management personnel would be capable of accessing the mainframe. If their chores were of significant scope they might have the capability of performing their own divisional functions with microcomputers. Lest I appear to be downgrading the micro, computer power has been so rigorously enhanced in the past decade that even some multimillion-dollar companies have entrusted all of their operations to microcomputers.

In the mid-tier, justly reserved for middle management, per se, is where activities such as purchasing, shipping and traffic dispatching, receiving, scheduling, manufacturing processing, and engineering are performed. Minicomputers or micros can be used in this functional area. Thus, the software packages that are available, or customized, could very well include computer graphics, scheduling, inventory control, CAD/CAM[1] traffic control routing, and the like.

239

At the very bottom tier are the microprocessors, microcomputers, and programmable logic controllers,[2] with specific software for particular applications and functions. This bottom tier of the computer environment includes shop-floor terminals, card-type data-entry systems, quality control, and materials-handling functions.

Database management systems

Computer numerical control (CNC) and direct numerical control (DNC) machining opened the way for computers in the manufacturing environment. After the CNC and DNC applications, it soon became apparent to manufacturing engineers that there were a lot of other uses for the computer in the factory. The technology that evolved permitted the control of industrial robots, the performance of quality control measurements, the compilation of QC statistics, and the control of automatic transfer machines and DNC equipment, all of which can be centrally and remotely controlled. Along with these developments, however, there arose an increasing necessity for a way in which all of these differing mechanisms could respond to a common computer language, which would be capable of providing the link (or *interface*, as it is called) between these machines. The engineering solution has resulted in the development of two protocols: one called MAP, and the other called TOP.

MAP is an acronym for Manufacturing Automation Protocol and is a term that originated at the General Motors Corporation in an effort to standardize computer communications both within its plantwide organization and with equipment suppliers and other concerned companies. It is a direct answer to the problems confronting the vast computer and control industries in linking the controls of one manufacturer's equipment with another's.

In fact, the prediction has been advanced that proprietary networks will no longer be used in new installations, and because of the MAP control system integration, the difficulty of combining various components from different manufacturers is now almost a dead issue.

In the past, up to half the cost of automating a new or existing facility was in integrating proprietary local area networks (LANs) in order to make them capable of communicating with each other. Therefore, eliminating this important element of cost makes it possible to undertake projects that would have had to have been abandoned because of the higher costs and complexities involved.

The TOP protocol, an acronym for Technical and Office Protocol, was developed by the Boeing Company engineers, and has many adherents in a number of major corporations. It links factory and office communications in much the same way as MAP.

Since the MAP and TOP protocols permit subfunction activities of a system to communicate, computer-integrated manufacturing (CIM) has become possible. This technological advance, therefore, has the capability of generating and storing useful data, and it is the management of this data that has become the foundation of the successful implementation of what is now known as a management information system (MIS) and its offshoot, the executive information system (EIS).

The MIS technology requires that a structural methodology exist to form the basis of support for CAD/CAM, group technology production, quality control, and a host of other activities that are required to monitor and direct plant performance. This support is known as *database management*.

It stands to reason that whenever an official action happens in the plant—viz. purchasing makes a buy, engineers write a procedure, or the manufacturing department turns a glob of iron into a piece part—information (data) is created. It is paradoxical that this data is almost as important as the part itself. Capturing this data effectively, therefore, involves the use of software programs having varying degrees of complexity and sophistication. By and large, software programs have been developed with a combination of modules that can be installed in stages so that implementation can be made in gradual phases, thus avoiding the mass confusion that might result if the complete package were to be installed in toto.

Thus, it is not always necessary for a user to install more modules than he or she thinks is desirable, or is capable of assimilating into the system. Nevertheless, the manufacturing software program should be planned as an integrated whole, or closed-loop system, that will permit management to plan for a prescribed number of units of production (based upon marketing data) and to follow the progress of every aspect of production from beginning to end product. Feedback from every department of the company should permit management to adjust the operation based upon real-time information that is part of the MIS program.

Achieving the full potential of manufacturing software packages requires a database software system between the packages and the mainframe computer that has the capability of orchestrating the complete system. Manufacturing processing software is so complex and interrelated with various processing components that enormous amounts of data must be read into the system on a constant basis in order to keep it running in a manner commensurate with the master schedule.

One of the principal advantages of a database software system is that it permits a specific element of data to be entered once, and immediately it appears in a file or module that requires this piece of information. As an example, if a change is made to a bill of materials, the database software will transmit this data to the purchasing department. If the item is a purchased part, it will be transmitted to the production scheduling group and/or quality control section. Therefore, the many advantages are obvious since it would be a waste of company funds if an obsolete part were to be purchased or manufactured, or if a parts listing was made invalid by the lack of change to its substantive content.

Although large-scale MIS systems are usually programs that are run on mainframe computers, microcomputer-based systems rely exclusively on their own standard files. The main disadvantage with this method is that each file must be updated separately. If this methodology were to be used in a mainframe computer, a standard file system would require an enormous amount of programming time for data-entry operations. Since the entries would have to be repeated many times over, the possibilities of operator error also would increase.

There are as many variations in database packages as there are vendors supplying this software, and their application modules might be somewhat different in the way in which they are programmed. As an example, capacity requirements planning (CRP) might be a subset of shop-floor control, or it might be a separate module entirely. Forecasting might be a separate module, or it might be a subset of the master schedule. Nevertheless, there is a logical sequence in which most database software packages can be installed. This sequence, quite frequently, determines in what order the programs should be implemented.

For the sake of this argument, it would probably be most advantageous to begin with the bill of materials because it directly affects all manufacturing phases. It is at this point that the manufacturing engineering department must determine how up to date its bills of materials really are. Prior to the installation of a good database management system, many companies find that their bills of materials require a rigorous updating. This is another advantage of introducing a database management system, since it forces the requirement for accurate data to be achieved by every affected manufacturing group. It has been stressed in chapter 6 that the bills of materials to be included in the database must be as accurate as it is possible to make them, and this fact is reemphasized at this juncture.

Subsequent to the bills of materials being entered into the computer would most likely be the buildup of the production schedule. In the main, this phase is accomplished by entering customer orders into the computer, which a software module converts into a production schedule, based on the minimum lot sizes of the manufacturing plant.

Although the production plan is thus approximated by the computer, it is only to be regarded as the base from which the production planner will revise and make adjustments predicated upon his or her experience and sound judgmental factors. Using the computer, it is possible to obtain a fairly representative idea of the capacity required to produce a particular manufacturing schedule.

This element of forecasting also permits potential shop bottlenecks to be spotted before they become actual problems in the plant. The advantage of using the computer in this manner enables make-or-buy decisions to be made, subcontracting to be provided for, and labor hiring requirements to be anticipated and planned for properly.

After all the bills of materials have been fed into the computer, they should be divided into their component parts, as indicated in chapter 6. In so doing, they can be summarized according to current and forecasted quantities and compared with quantities known to be in inventory and on order. In this manner an operator can be alerted to make any necessary adjustments.

With comprehensive programming, the computer will prepare purchase orders and revise schedules to conform to any adjustments in quantities and changes in machine tool loadings. The MRP systematic review of quantities can be made as often as required—sometimes on a weekly basis, but more often than not, on a monthly schedule.

As indicated in materials requirements planning and MRPII of chapter 6, another advantage of this methodology is that one can take a given piece part and determine which bills of materials list the part. Therefore, whenever an engineering change is made, the part can be tracked and noted as to the part, subassembly, or assembly in which the part is used.

This inherent capability of tracking parts indicates that the system has the power to generate cost data for each part and can compile labor, material, and overhead costs.

The standard costs derived for each subassembly and assembly as an output of the system provides a very effective mechanism for managing the plant.

Operating statistics

The materials management department is very closely associated with all of the operating departments of the plant, as indicated in everything that has been written in this text thus far. Materials management, therefore, generates as well as receives operating data. It also forms part of the MIS network.

An important link in the data network is the part of the system that tells exactly what is happening on the shop floor. This I/O (input/output) data describes what a particular machine tool or processing element has produced—i.e., in the nature of historical data—and what work awaits completion. This is where the need for shop-floor terminals that are integrated into the computer network is so essential to control.

The work that arrives at the work center or the machine is recorded by means of employee badges used at the computer terminal and compared with both the estimated time of completion of fabrication and the actual time the task was completed. In this manner, it is possible to monitor the work flow through the plant in a systematic fashion. Computer programming is developed or packaged programs are modified and adapted to the operating idiosyncrasies of the plant to make possible real-time comparisons between what is estimated to be achieved and what has actually been produced. Developing problems in this real-time system can be spotted rapidly and sorted out.

This manner of exerting an immediate corrective action is a way of indicating to the production worker that management has a handle on things and is capable of assuming the leadership role whenever it is necessary and required. An alert management gains the respect of employees, promotes good morale, and improves productivity in the process.

Another advantage of shop-floor data collection is that it permits standard hours to be revised and updated whenever there is reason to believe it is necessary. The collection of historical data for engineered standards is another management tool that justifies the use of computerized shop-floor control.

Computerized shop-floor control also permits the issuance of work orders, as well as engineering change orders. In this connection it is important that each production line foreman have a priority list of jobs. This list indicates that the right jobs are being worked on at the right time, and in the correct sequence. The line foreman can receive orders by means of a printed document or through the CRT display on the shop-floor terminal. If the terminal is used, the necessity for paperwork is eliminated and the work orders and the priority listing can be displayed on the tube.

The information required by the line foreman or shop supervisor translates into the following:

- Jobs to be done at each work center
- The job priority based on the date the job is to be completed
- A list of jobs to be received at the work center upon completion of the current tasks

The information transmitted to each shop foreman reflects the status of all jobs in the plant, with accurate locations of all parts and components. It also determines the priorities required to complete the assembly, or end product.

Production status reporting

As an example of an actual assembly line tracking and warehouse control system, there is the case in which a major New England manufacturer of computer printers desired to provide timely and accurate production status reporting in order to improve manufacturing productivity. The manufacturer also wanted to improve management of the warehousing and shipping activities in order that order picking and shipping documentation could be performed more effectively.

In implementation of a systems approach, management decided to use a WSSI/ Wand 2S and IBM Series/1 to collect data from bar-coded labels applied to printer assemblies, work orders, carton labels, and packing lists. The company uses scanning devices positioned at crucial points along the manufacturing production line. In addition, printer-assembly components are received from suppliers with bar-coded labels that identify each part. As each assembly completes the various phases in its fabrication, the identification labels are scanned prior to the assembly moving on to the next stage. Management reports are produced on the shop floor to provide the production control department with a real-time status of work flow and work-in-process inventory, together with a daily summary of units completed and released to the warehouse.

In the warehouse, on-line lightpen scanners are used to record the location where each printer is stored. To fill customer orders, a picking list with a bar-coded ID number is transmitted from the IBM Series/1 to a printer in the distribution center office. Products are selected by model number and prepared for shipment, then a portable scanner reads the picking list ID number and the bar-coded serial numbers on each carton for that particular order. This data is transmitted back to the host computer, where the shipping documentation is prepared and the warehouse stock locations and inventories are updated.

Time, attendance, and labor data collection

This system was designed to provide timely and accurate production status reporting—including the identification of active and inactive jobs at each work center, together with scheduled and nonscheduled operations in progress—and to locate all manufactured parts in the shop (location status). Another requirement was to provide shop supervisors with an accurate way to reconcile all direct and indirect labor reported for each employee. The management also wanted to provide a data collection system that would eliminate the time-consuming process of calculating and key-punching all the labor and clock-hour cards.

The system implementation was based upon the use of an IBM Series/1 and WSSI/ Wand 2S, just as recorded for the assembly line tracking system just described. The time, attendance, and labor data are collected by means of bar-coded badges, templates used in the production process, and data forms. The CRTs in the system are used for status inquiries and for the maintenance of the activity and maintenance files.

All machine-shop employees enter time, attendance, and labor-related information at work center reporting stations. Each station is equipped with a 32-character display, a light-pen wand, and a slot reader. Also used are bar-coded employee badges and templates with all of the functions and activities that will take place at the work center. All data is collected in a real-time continuous mode on the Series/1 by means of Wand 2S.

For time and attendance functions, the employee identifies him or herself by passing his or her badge through a slot reader. The employee clocks in and out at the beginning and end of his or her work shift and at the start and finish of a meal.

For all labor-related functions, the system guides the employee through the data-collection process by displaying visual messages or prompts on the scanning terminal. On-line editing is performed on the entered data against a database. The user-friendly interchange between the system and the user ensures timely and accurate data collection.

The product is routed from one work center to another based on instructions provided by a manufacturing order form that travels with the product. Orders are printed from the Series/1 using data transmitted from an IBM 3083 production scheduling system.

Shop supervisors can use the CRTs to make inquiries and to receive current status reports on jobs in the shop, and can review all employee activity reflected in the job reports. Supervisors can enter data in these terminals to add to the data-collection process. All time, attendance, and labor data collection by the Series/1 from bar-code readers and CRTs is periodically uploaded to the host computer as transactions to be processed. The host recalculates the production schedule, then downloads the revised schedule to the Series/1.

In this system, the cumbersome manual reporting of inaccurate labor data is replaced by current and accurate data collection by the work force. The use of time cards and labor cards is eliminated, thus reducing the personnel time required to process payroll and labor-distribution information. The bar-code data collection system ensures a high level of accuracy and data integrity, and reduces the downtime and effort required. CRT inquiries by shop supervisors keep them on top of employee activities and time and attendance information. CRTs also provide supervisors with up-to-the-minute information and hardcopy reports of work orders and manufactured parts on the shop floor, as well as jobs scheduled at each work center. Finally, the system provides accurate and effective lead-time information for capacity planning and production scheduling.

The system uses the CODE 39 bar-code design and the following hardware:

- 1 IBM Series/1, Model 4955F with 512K of memory
- 1 64 MB disk
- 1 .5 MB diskette
- 1 IBM 4978 CRT terminal
- 10 IBM 3101 CRT terminals
- 9 Intermec nondisplay bar-code readers with light pen
- 9 Intermec bar-code readers with visual display and light pens
- 6 Intermec slot readers with light pen
- 1 Intermec bar-code label printer

- 1 Printronix graphics printer
- 1 Intermec concentrator

The software used with this system is similar to that in the tracking system:

- WSSI/Wand 2S
- EDX operations and utilities
- PXS

Notes

1. CAD is an acronym for computer-aided design, a system that describes, in industrial applications, the more demanding and complex preparation of drawings and schematics. In these applications, the engineer or technician prepares a very detailed drawing on-line using a variety of interactional devices and programming techniques. Methods are required for duplicating basic figures; for achieving the exact size and placement of components; for making lines that vary in length, width, or angle to previously defined lines; for satisfying varying geometric and topological constraints among the components of the drawing; and the like. In addition to a pictorial datum or structure that defines where all the components fit on the picture (and that specifies their geometric composition), an application base is required to describe the electrical, mechanical, and other significant properties of the components. This should be in a form suitable for access and manipulation by the analysis program. The database has a further requirement in that it must be capable of being edited, in addition to being accessible to the interactive user.

   CAM is an acronym for computer-aided manufacturing, which is the use of computers and digital technology to generate manufacturing data. Data drawn from a CAD/CAM database can assist in controlling, or can control, part or all of a manufacturing process. This includes the command of numerical controlled machine tools and machining work centers, computer-assisted process planning, parts programming, robotics, and the command of programmable logic controllers. CAM can also control production scheduling and can assist in manufacturing engineering, facilities planning and engineering, industrial engineering, materials handling, and quality control. CAM can be used in many ways to assist in coordinating and operating the entire manufacturing facility.

2. Programmable logic controller (PLC) is in a class of microcomputers that have been optimized for special applications and can perform their functions faster and more efficiently than other types of general-purpose computers.

17

Using materials management to increase productivity

Productivity

American commerce and industry have thrived through the years, spurred on by a fierce, competitive spirit. Most of the time the consumer profits from this keen competition among companies in the business community; however, an unfortunate few in the manufacturing, distribution, or retail trade, who are submarginal producers, sometimes fail owing to a plethora of overwhelming inefficiencies. One of the most common of these defects is undercapitalization; the other is that of poor materials management practices, not the least among them is poor materials handling.

The trade literature is full of worthy examples of good materials management, good materials-handling practice, and new equipment all bent on improving productivity. Reading all the periodical literature that is available and buying new equipment might help some companies improve their productive effort; however, it is in putting together all the elements of a system in a common-sense way that will affect corporate productivity and profitability by the greatest amount.

As competition increases in an industry—on a global scale because the world is getting smaller—it is the relative efficiency of materials flow that, in the final analysis, helps achieve the least total cost of throughput and gives a sharp competitive edge to an organization. As an illustration, in the manufacturing area, machine tool builders supply the same highly automated, high-speed and feed equipment to you as they do to your competition. In warehousing and physical distribution, your competitor can employ the self-same powered industrial trucks or automated, high-rise, automatic storage and retrieval system (AS/RS) as you do. In order to stay ahead of competition, or in some instances just to stay even, you might have to revamp your thinking on planning, production, and materials control. After you have eliminated mechanical equipment and mechanization from your equation, you are left with one remaining area that is the essence of productivity improvement, and that is the concept of applying a systems approach to every facet of your operation.

Productive effectiveness of the highest order requires the following four ingredients:

- Quantity
- Quality
- Place
- Time

MRPII (chapter 6) demonstrated how it is possible to have the right quantity of materials at the workstation, but it did not say anything about quality. Therefore, it is up to QC and purchasing to bring materials up to the level required in order to improve productivity by eliminating defective materials before they reach the worker, and by minimizing defects in fabricated parts. The right place for the materials to be fabricated is in the plant prior to the time it is needed and at the workstation when it is required. The end product has to be finished on schedule, in the distribution cycle as required, and in the customer's hands at the time delivery is promised.

Materials flow

Materials management has the responsibility of improving materials flow. Therefore, eliminating machine operator downtime (due to delayed materials delivery to the workstation) is one way of achieving cost savings and improving employee morale, especially when workers are paid a premium based on good parts produced.

Indirect to direct labor ratios

In many industry segments, the increasing complexity and sophistication of equipment and machine tools, in general, has tended to increase the number of indirect labor workers in contrast to direct labor because of the service and maintenance requirements of the equipment. Since the machines are capable of producing more, with fewer direct labor workers, the number of direct workers has tended to decrease proportionately. Thus, the task faced by materials management is to systematically analyze material handling in relation to these automated machine systems and to minimize unnecessary handling (indirect) labor so that the productivity gains achieved by the faster and more complex machines is not lost.

Reducing handling damage

Any movement of materials, whether from supplier to plant, from plant to plant (i.e., interplant), or in-plant, has a tendency to increase the level of damage that occurs to the product being handled. Therefore, since scrap and rework can be costly, every means should be used to minimize losses caused by these material movements.

One way to hold losses to a minimum, of course, is to train handlers in the correct manner in which material should be transported. Another way is to make sure that you have provided the right equipment and containers for the materials in order to prevent damage. Also, in relation to containers is packaging and dunnage; sometimes it is necessary to provide good cushioning materials and well-fitting Styrofoam forms to transport fragile parts or parts with fine surface finishes. In the latter instance, parts with polished surfaces should be cleaned thoroughly before being transported, simply because of the abrasive effect of dirt and machine chips that can scratch and mar the finishes.

Although the foregoing has discussed some of the preventive measures to be taken to reduce materials-handling damage, a good damage-prevention program needs accurate data in order to fine-tune performance. Therefore, one of the first steps to be taken is to get handlers to fill out damage reports when such incidents occur. These reports will enable the materials management department to get a perspective on the causes of damage and enable a project manager to gauge the scope of the problem.

Selecting one individual in the materials management department to stay on top of the problem accomplishes two things: it helps focus attention on the seriousness of the problem, and it ensures followup when required. Another psychological factor is that having handlers (who, in general, detest paperwork) fill out damage reports will help prevent further damage.

A damage report form should contain the following information:

- Date
- Time
- Handler's name and unit
- Badge number
- Kind of damage
- Quantity (number of pieces)
- Cause of damage
- Signature of handler

If the handler hasn't filled out a damage report prior to the receipt of merchandise by the recipient, then the line foreman should seek out the materials control supervisor, who will, in turn, obtain the completed damage report from the handler. Thus, by compiling information from the periodic review of damage reports, the material management's project manager will be able to focus training and other, even disciplinary measures, to mitigate the problem of damage due to materials handling.

Maximizing space utilization

Since factory and warehousing space are increasing in cost with the upward inflationary spiral, the materials manager must look hard and fast at one of the more obvious methods for saving the company's funds and increasing the productivity and profitability (in a passive sense) of storing materials, both on the production line and in the physical distribution center. In the long run, the misuse of space is even more costly when viewed from the standpoint of the dislocations that occur when a move is made or when a plant expansion takes place because the plant has need of more space in which to operate.

In our prior discussion of facility planning, it was suggested that the materials-handling engineer should have a direct input in the planning stages of all new projects that dealt with either new or expanded plant space. There is little question that materials handling forms a crucial part of layout planning, but of equal importance is the interdependency of both production scheduling and inventory control with materials handling.

From the "systems" standpoint, lot size and the scheduling function have a bearing on the use of space in the plant, especially on production lines, and this question of quantity should be analyzed (optimized) by an industrial engineering study. The inventory control function is affected by the way in which work-in-process is stored and handled. Thus, to neglect one of the three areas—i.e., materials handling, production scheduling, and inventory control—is to skew the balance required to maximize the best utilization of plant space.

Reducing the accident rate and the severity of injury

Although the discussion in this chapter concerns productivity, it cannot disregard the factor of workplace safety. Motor vehicle and home accidents are much more frequent than accidents that happen in industrial surroundings. Perhaps one of the main reasons for the lower number of industrial accidents is industry's emphasis on safety, based in the main on the realization that accidents are costly in terms of both lost production and the higher rates that must be paid for workmen's compensation insurance.

Thus, in avoiding both lost production and higher compensation insurance rates, every company can contribute to the cause of productivity enhancement. Therefore, if your company does not now have a safety committee, it should hasten to initiate one and hire or train a safety engineer to guide a sound company safety program. A good thumbnail review of materials handling safety and training can be found in chapter 15 of my text, *Materials Handling: Principles and Practice* (Van Nostrand Reinhold, 1984).

Profit centers

Profit centers, like profit sharing, have their proponents and opponents. In the large multiplant companies, some managements are of the opinion that having each plant as a profit center makes it possible to spur a competitive spirit among the plants, which will lead to lower costs and higher profits. Other management groups feel that placing too much emphasis on plant profit centers might have a divisive effect on overall company operations since one group might feel compelled to protect its turf against another. The competitive spirit might get out of hand, and the rivalries that are engendered might do more harm than good.

Carried to its extreme, the profit center concept has even been tried in departments within a company. No one knows how effective or disruptive this might be since the intangible effects of this management philosophy are so hard to measure that what we have, mainly, is a great deal of philosophic discussion and "white papers" that, in the final analysis, are hardly conclusive.

The purpose of this discussion then is to examine the materials management department as a profit center in contrast to its consideration as a service function. Materials management functions have generally been regarded as services, just as plant maintenance is looked upon to help other profit centers within a company. Neither maintenance nor materials management has ever been regarded as a true profit center, particularly because their contribution to profit has been too indirect and difficult to measure.

Nevertheless, every materials manager can do a better job of supervising the activities of his or her department and improve performance by applying profit-center techniques. Since the purpose of the profit center is to generate larger profits, it is easy to understand why many of the larger companies tend to break up manufacturing into such centers, and following in this line of reasoning sales and engineering might also be considered candidates. Certainly, with the number-crunching capabilities of most computer systems, performance summaries should be able to be made easier than in the past.

A profit center is considered by the accounting department, almost, but not quite like a separate business. In other words, the profit generated by the center is a paper profit—real in terms of money and plant closings but imaginary in terms of overall company profits. As an example, an airplane assembly plant might be a separate profit center, buying its engines and other component parts from suppliers. It assembles these components into airplanes, which it can then sell to its sales department. The sales group sells to its customer airlines. The assembly plant's profit is the difference between its cost and the selling price to the sales organization.

The aircraft company's sales organization also might, in this context, be considered a profit center since it buys planes from the assembly plant and resells them to its customers. Thus, it buys materials—in this case planes—from the assembly plant, adds value to them through advertising and sales expense, sales promotional costs, and transportation, then ultimately resells them to the using airlines at a profit.

To parallel this illustration of a profit center, it is possible to consider materials management in much the same light. It buys materials (through its purchasing department) from others, adds value to them in the form of distribution (materials control), then sells them to the manufacturing department. The difference between the price at which the material is sold (transferred) to manufacturing control and materials management overhead (distribution) represents the (fictitious) profit accruing to materials management. Of course, this is all a so-called paper profit and bears no relation to the profit generated by the company, since what this concerns primarily is a way to measure the effectiveness of the materials management organization.

There are three basic criteria that every profit center must measure up to:

- Value must be added in some way.
- Capital must be employed productively.
- Costs are incurred.

A materials management group in any manufacturing or nonmanufacturing organization can meet these criteria as easily as the captive airplane assembly plant of the aircraft company just cited. Notwithstanding the argument presented here, I do not know of any materials management organizations that are actually, for accounting purposes, considered profit centers. A sufficient reason is that there might not be any top management executives who are at all familiar with this concept, and if any exist they might not think it advantageous to the objectives of their companies. Also, the concept of materials management as a profit center implies the existence of a completely independent materials management organization, and this thought might not always receive a warm welcome in companies where materials management has been subordinated to manufacturing (usually), finance, or some other top-level department.

Thus it is that top management might not have any real enthusiasm for placing its materials manager in charge of his or her own profit center for the sake of enlarging his or her authority in the company. However, top management generally can be encouraged to permit the materials management organization to set up its own profit-and-loss balance sheet if it will assist in contributing to corporate profit.

18

The effects of
product mix on
materials management

Product mix and profitability

In many manufacturing companies, the cost of materials handling represents from 20 to 35 percent of the cost of the product. For some agricultural products and foodstuffs the cost figure might be even higher. Since materials handling is such a large element of cost, one of the primary concerns of materials management is to try to eliminate or minimize the handling of materials wherever possible. Even in some of the more mechanized companies, there is always an opportunity for improving and eliminating one or more of the many materials-handling activities that exist.

Since materials handling is concerned with the movement of materials, every movement has to have a pick-up element and a set-down, and most of the time there is a transportation distance between these two elements or points. Thus, if the pick-up cost and the set-down costs are relatively fixed, then the sole variable that remains for manipulation is the transportation distance between the two points. Quite elementary, naturally, but a factor that is often lost sight of, even in the most modern of manufacturing plants. Therefore, as small a factor as this might seem, as the transportation distance increases, so does the cost per unit of product handled.

In this context, there are three basic characteristics of materials handling:

- Picking up the load
- Transporting the load
- Setting down the load

In addition, there are two opposing elements of cost that eventually enter every materials-handling problem:

- Product mix
- Load size

In the province of materials management, the *product mix* describes the number of different sizes, shapes, and types of products that must be handled. Therefore, as the product mix increases, the cost of handling increases because of the difficulty of handling products of several sizes. As an example, if 21-foot-long pipes, rods, or bars, and cartons, steel drums, and varying-sized pallet loads are all received across the same receiving platform, the different methods and types of handling equipment used to transport these diverse items will add to the complexity and, thus, the cost of handling. On the other hand, if only corrugated carton shipping containers of a certain size are received, then the handling problem is relatively simple and materials-handling equipment can be standardized—perhaps even employing the use of conveyors—in order to keep the cost per unit handled at a relatively low level.

Thus, if it is possible to keep the type of product mix low, then it will be possible to keep handling costs down. This is not always possible in most manufacturing and service organizations, however; and the best that can be done in most situations is to be prepared with the proper handling equipment and methods, preparatory to receiving certain types of materials that are not customarily, or frequently, handled by the company.

Another of the opposing elements in materials handling practice is that of load size. As the load size increases or decreases, costs will vary, depending on several factors. As an example, if the unit load size increases, handling costs will invariably decrease, since the per-unit cost will decrease, just as it is less costly to transport a pallet load of bags of cement than to carry one bag at a time. Also, if the size of the company will justify the method, it would be more economical to handle cement in bulk tank cars than to pack the cement in bags, then palletize the bags.

The point is that, as the load size increases, the unit cost of handling decreases, provided that the volume of materials to be handled justifies the cost of the equipment required to do the job. For some smaller investments the R.O.I. might be six months to a year. On larger mechanization projects the return might require two or more years.

Diversification philosophies

During the last surge in interest rates, when some of the smaller companies were paying as much as 18 percent or more per annum for operating capital, some submarginal producers were forced to shut their doors permanently. This economic aberration, hopefully, is a thing of the past, but the lesson to be learned is that there is a potential danger in having only a few products in the company line, especially if the items are high priced and customer's purchases must be financed with expensive dollars.

In the previous section, the discussion centered upon the materials management department's concern with the product mix of items to be handled. It was concluded that the product mix of incoming merchandise—receivables—increased the cost of handling; thus, preparatory measures had to be taken to mitigate this problem. In a diversification philosophy, a certain amount of end-product mix is not only necessary, but desirable, for heavy-equipment producers if the possibility exists that interest rates might rise again.

It is for this reason that materials management should opt for the continued study by the marketing department in the area of new product development. The contribution by materials management is in its purchasing department function, which can be the very eyes and ears of top management as its personnel studies the trade and periodical literature, and discusses with its many suppliers what is happening in the various segments of industry.

19

Applying the systems approach in materials management

Labor

The computer has extended the scope and depth of the materials management function and has permitted better and tighter control of inventory and other materials management activities. Because of the capability of the computer to handle vast amounts of data, the productivity of materials management personnel has increased enormously in the past decade. In the modern materials management department, strictly clerical jobs are largely a thing of the past, and the emphasis is now on highly trained, college-level personnel with degrees in mathematics and economics.

Today's highly sophisticated inventory control formulas are being used and developed anew by these mathematically skilled technicians. Also, as factories and offices introduce more complex, automated systems, the need for economists who can prepare meticulously precise forecasts of future demand becomes more necessary, even as computers help them in analyzing the masses of data that are currently available.

All kinds of factory automation require personnel with manufacturing engineering backgrounds in the materials management department in order to cope with tighter quality standards, close and concise factory production schedules, and the more complex manufacturing methods that automated processing entails.

It is no longer possible, except in the very smallest of companies, to promote a man or a woman to the position of buyer or purchasing agent simply because of adequate or competent performance in the storeroom and a pleasing personality. In today's material management organization, competent people with college backgrounds are not only desirable, but their educational backgrounds must include college training. In some of the larger companies, advanced degrees are required.

The latest methodologies concerning materials management view this function as a subset operating within an overall closed-loop, management operating system and obtaining feedback from a MIS (see chapter 16). The implications for materials management personnel (or candidates for this department) are that the traditional company organization—in which management is viewed as a combination of highly specialized functions, such as engineering, manufacturing, marketing, finance, purchasing, etc. (as shown in FIG. 1-1)—is no longer a valid thesis. The traditional approach is relatively static and organizationally rigid. The systems approach applied to manufacturing and nonmanufacturing organizations alike has changed this view, mainly through the extensive and intensive application of computer power.

The opportunity for employees in this new arena are many and challenging because the systems approach is essentially dynamic, and change is a constant. Functions that were viewed as static and rigidly defined in the traditional, organizational structure have become extremely fluid. In the systems approach, each functional activity has the capability and necessity of interacting with every other activity. Instead of specialists in each field, the tendency is to develop generalists who are constantly involved in the activities of others.

Thus, the materials management organization, in which all materials movements and activities are grouped together under a common manager, is ideally suited to a systems approach to management. It harnesses the power of an MIS and interacts with other subsystems in the total organization complex.

Capital investment

Materials management as a methodology is firmly entrenched in Japan and Europe. The United States has been slow to adopt this concept, perhaps because manufacturing industries have a greater importance in other countries by contributing more to their gross national products than here in the United States.

For reasons that defy economic logic in our country, materials management is more commonly associated with manufacturing than any other segment of the economy. It wasn't until well into the midpoint of the 1900s that manufacturing ceased to become the most important segment of the economy, and manufacturing companies were only then becoming aware of the existence of materials management. Nonmanufacturing enterprises ignored it almost completely.

Since the end of World War II, however, manufacturing has been declining in importance, supplanted by services in both the public and private sectors of the economy. Nonmanufacturing has been contributing an ever-increasing share of the national output. With this upward thrust has come the realization that materials management is as important a concept to producers of services as it is to producers of commodities.

In this respect, the railroad companies were among the first in the nonmanufacturing segment of the economy to recognize materials management as a separate entity within their organization. Since this time more nonprofit and public service groups have decided that they have a vested interest in the movement of materials, and there are quite a few such institutions with dedicated materials management organizations.

Value added by manufacturing

The physical process of adding value by manufacturing is relatively easy to understand. A press operator at a stamping or punch press places a blank piece of metal or a strip of steel onto a die between two huge jaws and presses fingers of both hands on two widely separated buttons. The jaws come together with a clang and a piece part drops into a container in back of the machine. A bunch of different stampings are welded or bolted together with some other components, and in short order an end product is assembled.

This is only one of many manufacturing activities, for in some companies the actual manufacturing might amount to little more than assembling components purchased from vendors. Nevertheless, it is quite easy to visualize the functions comprising the adding of value in manufacturing.

Value added by distribution

Adding value by distribution is somewhat more difficult to visualize than the manufacturing concept. Distribution is the process by which the end product arrives in the hands of the ultimate consumer. The distribution process begins after the last manufacturing operation is completed. At this point what usually happens is that the product is packaged, stored, shipped, and sold to a customer. There are some slight variations in this chain of events, but more or less the process is essentially as routine as it sounds, although value is also added in a less direct manner.

As an example, in the course of distribution, the material or end product might be stored in a warehouse or distribution center. In this fashion, the warehousing cost adds value, as does the value of money—i.e., interest on the money tied up in inventory. Also, if the merchandise is subject to damage or pilferage, then this loss also becomes part of the value added by distribution to the products that ultimately reach the consumer.

Since manufacturing cannot exist without distribution, the two activities are inter-related; however, much of the value that is added by distribution is found in organizations that do no manufacturing whatsoever, such as discount store chains that buy radios from RCA for $100 and sell them for $200, adding $100 worth of value by distributing the RCA end product.

The employment of capital

In today's economy, as it has ever been in a capitalist economy, it is impossible to produce anything without at least a modicum of investment capital. An investment in machinery, equipment, buildings, and in-process inventories is absolutely essential if value is to be added by manufacturing.

Distribution functions also require capital, and sometimes this amounts to a sizable investment. For example, the discount store chain owner needs buildings, inventory, and other capital goods with which to operate. The profit that is earned on sales represents the return that this capital must earn to be productive.

Capital investment in inventory that does not contribute to profitability requires reinvestment elsewhere. Thus, one of the major objectives of materials management is to identify inventory that is not working to add value by distribution and remove it as quickly as possible, in any way that will recoup at least part of the investment.

Use of facilities and equipment

With MIS availability, the materials management department can make optimum use of facilities and equipment in the following manner:

Facilities

By holding inventories to a minimum quantity at each workstation, it might be possible to free up expensive manufacturing space that is presently being used for work in process. Through purchasing's supplier network, materials management can weed out slow and unreliable suppliers and strive to obtain just-in-time supply support.

Equipment

- As an adjunct to the MIS, materials management can evaluate the benefits of electronic data interchange (EDI) use by purchasing and traffic as a company-to-company exchange of common business data, as found in invoices, purchase orders, shipping notices, and the like. EDI is the keyless transmission of data between computers in different companies in which data is initially entered into the computer by an automatic identification methodology, such as bar code, magnetic stripe, OCR, and the like, although it also can be keyboarded. Since the communication is between companies, a third-party EDI network is frequently employed as a translator of various computer formats and as a post office to store information until it is transmitted directly to the receiving computer for processing.

    When a bill of lading is electronically transmitted to a warehouse, the arriving merchandise having a bar-coded label is scanned and recognized by the receiving department. EDI reduces lead times and improves the accuracy of transmitted data.
- Historical data obtained through the MIS on shop-floor management can establish operating standards for materials control operators, and can improve work flow throughout the manufacturing plant.
- RF/DC (radio frequency data communication) terminals also can be used. These devices are hand-held or mounted on forklift trucks, or other mobile materials-handling equipment. For use by materials control, the RF/DC allows shipping, receiving, order-picking, storage and retrieval, and other instructions to be transmitted directly to and from operators at each terminal and a host computer.
- Another possibility to explore is the use of RF identification (RF/ID), which is similar to having radio transmitters and receivers and is used in the following manner: electronic tags, or *labels*, are programmed with specific information and attached to items that need to be tracked or identified, such as unit loads of materials; vehicles, and containers. Strategically located antenna sensors read the information as the tagged objects pass by.

20

Implementing the materials management concept

Preparing the climate

In most instances, the chief executive of a company would probably be far too busy or disinterested to be concerned about the need for a materials management organization in the company, were it not for some problems in supply or missed shipping schedules that have caused customer complaints, and the like. Very few chief executives want to rudely shake the existing organizational structure unless they are very young, very brash, or well heeled. Sometimes the political in-fighting in the executive suite can become so bitter and intense that some type of change is inevitable.

Support for the change to a materials management type of organization often comes from the person or persons who consider themselves the best possible candidates for the job of materials manager. Unfortunately, making as sweeping a change as an integrated materials management department requires a tremendous amount of corporate clout, and many candidates or proponents gauge the political temperature and decide to put together their organizations in small bite-sized chunks. This situation might occur even when the proponent gains the ear of some senior official, say on the vice-presidential level, does his homework, and prepares a brilliant dissertation on the many advantages that will accrue to the company following adoption of the concept.

Fortunately, materials management can be installed in small modules, and this is a convincing way of selling the idea—giving a timetable that inspires confidence since it does not anticipate sweeping changes or major disruptions to the way the plant presently operates.

Organizing for change

Although there is only one CEO in each company, today's business enterprises are so complex with tax considerations so abstruse, that most business decisions are arrived at by groups of individuals. It is only in very small, closely held companies that business decisions are made by only one or two people, and then not without the assistance of legal experts and tax advisors. Thus, it is up to the materials management proponent to sell management on the soundness, economic benefits, and other advantages of the new organization.

There are a number of excellent texts on the subject of presentations; therefore, only the highlights will be included here, to serve merely as a checklist:

1. Organization
   - Establish the objectives of the presentation.
   - Examine the way your audience thinks.
   - Try to find out their interests, motivations, and expectations.
   - Know who will attend; make certain they are the "right" people.
2. Develop a Control Plan
   - Make certain that you know what the "highs" and "lows" are and plan around them.
   - Be sure your presentation ends on a "high" note.
   - Try to obtain as much time as you'll need.
   - Try to pick the best time for the meeting.
   - Try to avoid times when there might be meeting conflicts for your most important listeners.
3. Develop a Complete Presentation
   - Overprepare; have all the bases covered.
   - Anticipate questions and problems.
   - Be as concise as possible; don't ramble.
   - Develop a management overview; include objectives, recommendations, and an action plan.
   - Distribute reports of the presentation a few days to a week before the meeting in order to give management an opportunity to digest the material to be presented.
4. What to Include in the Presentation
   - Project objectives (list them in terms of their importance).
   - Various parameters involved.
   - Requirements.
   - Opportunities.
   - Benefits, such as return on investment.
   - Recommendations with possible alternative solutions.
   - Be conservative in all estimates, especially of expected results.
   - If possible, list similar installations as examples.
   - Description of the system or systems.
     - Operational capacity.
     - First cost and operating costs.

○ Flexibility.
○ Expandability and vice versa.
○ Economics of a parallel operation during installation.
5. Summary
   • Give management the time to make a decision.
   • Summarize the meeting.
   • Don't let the meeting end before obtaining specific dates for the next meeting and the names of personnel to continue on the project. Ask for the commitment and involvement of management.

Implementation

Although most top management groups can get very tough when the occasion warrants, in the main, they are usually somewhat sensitive to the feelings of their subordinates. Very few organizations like "boat-rockers," and the resistance to change is almost a fact of life in almost every company. Look how long it took to topple the Berlin Wall!

For this reason, major changes are almost always postponed, and interminable meetings are held, as discussion succeeds discussion. In some instances, people will resign or retire, creating an opening, so that at least some action can be taken.

Often, organizational changes are made in such a manner that everyone winds up with at least as good a job as he or she had prior to the change, and with a little more money as a sweetener. Moving a transplanted purchasing manager to a larger division, a better office, etc., can do much to implement the required changes. Other changes might be made by attrition as employees leave the company, are transferred to other plants, and the like. Sometimes layoffs happen, and during the recession changes can be made unobtrusively.

Personnel changes should almost never be made in such a fashion that employees are penalized. Even very small monetary increases can create a better feeling for the changes that will benefit the company to such a large degree.

As far as the actual introduction of methodology is concerned, you are advised to review this text, select those portions of the materials management schema that will best suit your company, and then proceed to take steps to ensure that a smooth transition will be made.

Appendix A

Automatic identification, hardware and software suppliers, and other peripherals

ACCU-SORT SYSTEMS
511 School House Rd.
Telford, PA 18969 – 9990
215 – 723 – 0981/800 – BAR – CODE

ALLEN-BRADLEY CO.
1201 S. Second St.
Milwaukee, WI 53204
414 – 382 – 2000

AVERY SOABAR PRODUCTS GROUP
7722 Dungan Rd.
Philadelphia, PA 19111
215 – 725 – 4700

W.H. BRADY CO.
2221 W. Canden Rd.
Milwaukee, WI 53209
414 – 351 – 6630

COMPUTER IDENTICS CORP.
5 Shawmut Rd.
Canton, MA 02021
617 – 821 – 0830/800 – 343 – 0846

COMPUTYPE, INC.
2285 W. County Rd. C
St. Paul, MN 55113
612 – 633 – 0633/800 – 328 – 0852

CONTROL MODULE, INC.
380 Enfield St.
Enfield, CT 06082
203 – 745 – 2433/800 – 722 – 6654

DATALOGIC, INC.
301 Gregson Dr.
Cary, NC 27511
919 – 481 – 1400

DIAGRAPH CORP.
3401 Rider Trail S.
St. Louis, MO 63045
314 – 739 – 1221

FAIRBANKS SCALES
711 E. St. Johnsbury Rd.
St. Johnsbury, VT 05819
802 – 748 – 5111

GENERAL ELECTRONIC SYSTEMS, INC.
914 SE 14 Place
Cape Coral, FL 33990
813 – 574 – 2313

GRAND RAPIDS LABEL CO.
2351 Oak Industrial Dr. NE
Grand Rapids, MI 49505
616 – 459 – 8134

HAND HELD PRODUCTS, INC.
8008 Corporate Center Dr.
Charlotte, NC 28226
704 – 541 – 1380

HEWLETT-PACKARD CO.
3000 Hanover St.
Palo Alto, CA 94303
408 – 434 – 7305

IBM CORPORATION
Old Orchard Rd.
Armonk, NY 10504
407 – 443 – 9344/800 – IBM – 6676

IDENTATRONICS, INC.
425 Lively Blvd.
Elk Grove Village, IL 60007
708 – 437 – 2654

IMTEC, INC.
One Imtec Ln.
P.O. Box 809
Bellows Falls, VT 05101
802 – 463 – 9502

INTERMEC
6001 36th Ave. W.
P.O. Box 4280
Everett, WA 98203 – 9280
206 – 348 – 2600

KEY-TRONIC CORP.
P.O. Box 14687, Mail Stop 142-D
Spokane, WA 99214 – 0687
913 – 345 – 0690

KIMBERLY-CLARK, BROWN-RIDGE
518 E. Water St.
P.O. Box 370
Troy, OH 45373 – 0370
513 – 339 – 0561

LOGISTICON, INC.
4001 Burton Dr.
Santa Clara, CA 95054
408 – 988 – 3811

LXE, INC.
303 Research Dr.
Norcross, GA 30092
404 – 447 – 4224

MANNESMANN TALLY CORP.
8301 S. 180th St.
Kent, WA 98032
206 – 251 – 5500

MARKEM CORP., SCANMARK DIVISION
150 Congress St.
Keene, NH 03431
603 – 352 – 1130

MONARCH MARKING SYSTEMS
One Kohnle Dr.
P.O. Box 608
Dayton, OH 45401
800 – 243 – 4015

MOORE BUSINESS FORMS
1205 Milwaukee Ave.
Glenview, IL 60025
708 – 480 – 3000

PRINTRONIX
17500 Cartwright Rd.
Irvine, CA 92713
714 – 863 – 1900/800 – 826 – 3874

RJS, INC.
140 E. Chestnut Ave.
Monrovia, CA 91016
818 – 357 – 9781

SECURITY TAG SYSTEMS, INC.
1615 118th Ave. N.
P.O. Box 23000
St. Petersburg, FL 33716
813 – 576 – 6399

SYMBOL TECHNOLOGIES, INC.
116 Willbur Place
Bohemia, NY 11716
516 – 563 – 2400

TEXAS INSTRUMENTS
34 Forest St.
Attleboro, MA 02703
508 – 699 – 1639

UARCO
W. County Line Rd.
Barrington, IL 60010
312 – 381 – 7000

WEBER MARKING SYSTEMS
711 W. Algonquin Rd.
Arlington Heights, IL 60005 – 4457
708 – 364 – 8500

WELCH ALLYN, INC.
Jordan Rd.
Skaneateles Falls, NY 13153 – 0187
315 – 685 – 8945

Appendix B

Occupational Safety and Health Administration (OSHA)
Subpart N— Materials Handling and Storage
Para. 1910.178
Powered Industrial Trucks

NOTE: The Williams-Steiger Act of 1970, which created OSHA, affects every employer of more than seven workers throughout the United States. The purpose of this appendix is to illustrate how the public law affects the materials management organization and its areas of responsibility with regard to powered industrial trucks.

Para. 1910.178 Powered industrial trucks

a. General requirements

1. This section applies to all of the mobile equipment used by materials handling practitioners.
2. All new, powered industrial trucks acquired after February 15, 1972 shall meet the design and construction requirements of ANSI – B56.1 – 1969.
3. Approved trucks must have identifying labels as indicated in ANSI – B56.1.
4. A very important note concerns the modifications to industrial powered trucks—any modification that may affect capacity or safety cannot be per-

formed by the user without the truck manufacturer's prior written approval. When modifications are made—for example, when a rotator attachment is added to the truck—a new label has to be added indicating the attachment and change in capacity of the truck.

5. If the truck is equipped with front-end attachments other than factory-installed attachments, the user must request that the truck be marked to identify the attachments and show the weight of the truck and attachment combination, and to indicate the capacity of the truck with the attachment as maximum elevation with a load that is centered on the attachment.

6. The user must ensure that all truck nameplates and markings are in place and maintained in a legible condition. This is one of the most abused of the regulations dealing with industrial powered trucks. I have found it almost impossible to comply totally with this regulation.

b. Designations

The truck user should be able to recognize the following symbols for powered industrial trucks:

D — diesel

DS — diesel with exhaust, fuel, electrical safeguards

DY — all DS features including temperature limitation features, and no electrical or ignition equipment

E — electric

ES — electric with spark arrestors and surface temperature limitations

EE — electric with completely enclosed electrical systems

EX — electric, explosion-proof

G — gas

GS — gas with exhaust, fuel, and electrical safeguards

LP — liquefied petroleum (propane)

LPS — liquefied petroleum with exhaust, fuel, and electrical safeguards

c. 2.i. Power-operated industrial trucks shall not be used in atmospheres containing hazardous concentrations of acetylene, butadiene, ethylene oxide, hydrogen, and so forth.

12.2.ii. Power-operated industrial trucks shall not be used in atmospheres containing hazardous concentrations of metal dust, including aluminum, magnesium, carbon black, coal or coke dust, except approved EX-rated trucks.

NOTE: If you have any doubt as to what trucks can be used in certain locations, refer to the *Federal Register*, chapter XVII, Occupational Safety and Health Administration, Subpart N, 1910.178, Table N-1 "Summary Table on Use of Industrial Trucks in Various Locations."

d. Converted industrial trucks

Trucks that were originally approved for the use of gasoline for fuel, when converted to LP gas can be used in those locations where G, GS, or LP and LPS trucks have been designated for use provided that the truck conforms to the GS and LPS requirements.

e. Safety guards

Overhead guards and load backrest extensions must be provided on all trucks where lifts higher than six feet are made.

f. Fuel handling and storage

1. The storage and handling of liquid fuels such as gasoline and diesel fuel shall be in accordance with NFPA Flammable and Combustible Liquids Code (NFPA No. 30-1969).
2. The storage and handling of LP gas fuel shall be in accordance with NFPA Storage and Handling of Liquefied Petroleum Gases (NFPA No. 58-1969).

g. Changing and charging storage batteries

1. Battery charging installations shall be located in areas designated for that purpose.
2. Facilities shall be provided for flushing and neutralizing spilled electrolyte, for fire protection, for protecting charging apparatus from damage by trucks, and for adequate ventilation for the dispersal of fumes from gassing batteries.
3. Battery racks shall be nonsparking.
4. A conveyor or hoist for changing batteries should be provided, depending on the type of battery and truck.
5. Reinstalled batteries shall be properly positioned and secured in the truck.
6. A carboy tilter or siphon shall be provided for handling the electrolyte.
7. When charging batteries (or mixing electrolyte), acid shall be poured into water, never the reverse.
8. Trucks shall be properly positioned with their brakes applied before attempting to change or charge batteries.
9. Keep vent caps in place when charging batteries. Make sure the vent caps work. Open the battery compartment cover to dissipate the heat.
10. No smoking in the charging area. Post the area accordingly.
11. No welding in charging area. Precautions should be taken to prevent open flames, sparks, or electric arcs in the battery charging areas.
12. Tools, wrenches, etc., should be kept away from the tops of uncovered batteries; otherwise serious, or fatal, short circuits may result.

h. Lighting for operating areas

1. Controlled lighting of adequate intensity should be provided in operating areas. (See American National Standard Practice for Industrial Light, A11.1-1965 [R.1970], which can be obtained at many public libraries.)

2. Where light levels are less than two lumens per square foot, auxiliary directional lighting shall be provided on the truck.

i. Control of noxious gases and fumes

Concentration levels of carbon monoxide (CO) gas created by powered industrial truck operations shall not exceed the levels specified in Para. 1910.1000 (Subpart Z) of OSHA. From (Subpart Z) 11910.1000:

$$\frac{Table\ Z\text{-}1}{CO\ =\ PPM\ 50^a\ 55\ mg/M^3\ ^b}$$

where a is parts of vapor or gas per million parts of contaminated air by volume at 25 °C and 760 mm Hg pressure, and b is approximate milligrams of particulate per cubic meter of air.

Any qualified industrial hygienist can measure these concentrations if you have any doubt as to whether or not you can comply with OSHA standards in particular areas of your plant.

j. Dockboards (Bridge Plates) [from 1910.30(a)]

Portable and powered dockboards shall be strong enough to carry the load imposed on them. They must have some means to anchor them to keep them from slipping. Portable dockboards must have handholds to lift them so they can be safely handled manually. Usually dockboards are positioned and placed at a car or truck spot by forklift truck. Dockboards of 20,000 lbs. or larger capacity cannot be handled manually because of their weight.

Powered dockboards have to be designed and constructed according to Commercial Standard CS202-56 (1961) "Industrial Lifts and Hinged Loading Ramps," published by the U.S. Department of Commerce.

k. Trucks and railroad cars

Positive protection must be provided to prevent railroad cars from being moved while dockboards or bridge plates are in position.

The same is true of truck loading and unloading operations with dockboards, since air lines and wheel chocks have to be in place and brakes set in order to prevent the truck from moving. In addition, if a semi has only the van trailer at the truck spot—that is, the tractor has been removed and the box is sitting on front landing wheels—then two jacks have to be placed under the nose of the van trailer to prevent overturning.

l. Operator training

Only trained and authorized operators shall be permitted to operate a powered industrial truck. Methods shall be developed to train operators in the safe operation of powered industrial trucks.

m. Truck operation

OSHA requires that operators of powered industrial trucks be knowledgeable concerning the safe operation of their equipment. It would be well for a plant that uses powered industrial trucks on a continuous basis to invest in a packaged training program if it does not have one of its own. Many of the larger forklift truck manufacturers can send you manuals on forklift truck operator training.

1. Trucks shall not be driven up to anyone standing in front of a bench or other fixed object.
2. No person shall be allowed to stand or pass under the elevated portion of any truck, whether loaded or empty.
3. Unauthorized personnel shall not be permitted to ride on powered industrial trucks. A safe place to ride shall be provided where riding of trucks is authorized.
4. The employer shall prohibit arms or legs from being placed between the uprights of the mast or outside the running lines of the truck.
5. i. When a powered industrial truck is left unattended, load-engaging means shall be fully lowered, controls shall be neutralized, power shall be shut off, and brakes set. Wheels shall be blocked if the truck is parked on an incline.
5. ii. A powered industrial truck is unattended when the operator is 25 feet or more from a vehicle that remains in his view, or whenever the operator leaves the vehicle and it is not in his view.
5. iii. When the operator of an industrial truck is dismounted and within 25 feet of the truck still in his view, the load-engaging means shall be fully lowered, controls neutralized, and the brakes set to prevent movement.
6. A safe distance shall be maintained from the edge of ramps or platforms while on any elevated dock, platform, or freight car. Trucks shall not be used for opening or closing freight doors.
7. Brakes shall be set and wheel blocks shall be in place to prevent movement of trucks, trailers, or railroad cars while loading or unloading. Fixed jacks might be necessary to support a semitrailer during loading or unloading when the trailer is not coupled to a tractor. The flooring of trucks, trailers, and railroad cars shall be checked for breaks and weakness before they are driven onto.
8. There shall be sufficient headroom under overhead installations, lights, pipes, sprinkler systems, etc.
9. An overhead guard shall be used as protection against falling objects. Note that an overhead guard is intended to offer protection from the impact of small packages, boxes, bagged material, etc., representative of the job application, but not to withstand the impact of a falling capacity load.
10. A load backrest extension shall be used whenever necessary to minimize the possibility that the load or part of it will fall rearward.
11. Only approved industrial trucks shall be used in hazardous locations.
12. Whenever a truck is equipped with vertical only, or vertical and horizontal, con-

trols elevatable with the lifting carriage or forks for lifting personnel, the following additional precautions shall be taken for protection of personnel being elevated.

i. A safety platform firmly secured to the lifting carriage and/or forks will be used.

12. ii. Means shall be provided whereby personnel on the platform can shut off power to the truck.

12. iii. Such protection from falling objects as indicated necessary by the operating conditions shall be provided.

13. Reserved   To date nothing further has been added to section 13, but it would be advisable to request the plant safety officer to check this out. If there is no plant safety officer in your organization, then a call to the local OSHA office listed in your telephone directory should bring you current information on this subject. You do not have to give your company name in order to get this information.

14. Fire aisles, access to stairways, and fire equipment shall be kept clear.

n. Traveling

1. All traffic regulations shall be observed, including authorized plant speed limits. I have established the following rules for manufacturing operations: 5 mph on main aisles with two-way traffic and 3 mph on side aisles and other aisles. A safe distance shall be maintained approximately three truck lengths from the truck ahead, and the truck shall be kept under control at all times.

2. The right of way shall be yielded to ambulances, fire trucks, or other vehicles in emergency situations.

3. Other trucks traveling in the same direction at intersections, blind spots, or other dangerous locations shall not be passed.

4. The driver shall be required to slow down and sound the horn to cross aisles and other locations where vision is obstructed. If the load being carried obstructs the forward view, the driver shall be required to travel with the load trailing.

5. Railroad tracks shall be crossed diagonally wherever possible. Parking closer than 8 feet from the center of railroad tracks is prohibited.

6. The driver shall be required to look in the direction of, and keep a clear view of, the path of travel.

7. Grades shall be ascended or descended slowly.

   i. When ascending or descending grades in excess of 10%, loaded trucks shall be driven with the load upgrade.

7. ii. Unloaded trucks should be operated on all grades with the load-engaging means downgrade.

7. iii. On all grades, the load and load-engaging means shall be tilted back if applicable, and raised only as far as necessary to clear the road surface.

8. Under all travel conditions the truck shall be operated at a speed that will permit it to be brought to a stop in a safe manner.

9. Stunt driving and horseplay shall not be permitted.

10. The driver shall be required to slow down for wet and slippery floors.

11. Dockboards or bridge plates shall be properly secured before they are driven

over. Dockboards or bridge plates shall be driven over carefully and slowly, and their rated capacity never exceeded.

12. Elevators shall be approached slowly, then entered squarely after the elevator car is properly leveled. Once on the elevator, the controls shall be neutralized, power shut off, and the brakes set.

13. Motorized hand trucks must enter elevators or other confined areas with the load end forward.

14. Running over loose objects on the roadway surface shall be avoided.

15. While turns are being negotiated, speed shall be reduced to a safe level by means of turning the hand steering wheel in a smooth, sweeping motion. Except when maneuvering at a very low speed, the hand steering wheel shall be turned at a moderate, even rate.

o. Loading

1. Only stable or safely arranged loads shall be handled. Caution shall be exercised when handling off-center loads that cannot be centered.

2. Only loads within the rated capacity of the truck shall be handled.

3. The long or high (including multiple-tiered) loads that may affect capacity shall be adjusted.

4. When attachments are used, particular care should be taken in securing, manipulating, positioning, and transporting the load. Trucks equipped with attachments shall be operated as partially loaded trucks when not handling a load.

5. A load-engaging means shall be placed under the load as far as possible. The mast shall be carefully tilted backward to stabilize the load.

6. Extreme care shall be used when tilting the load forward or backward, particularly when high tiering. Tilting forward with load-engaging means elevated shall be prohibited except to pick up a load. An elevated load shall not be tilted forward except when the load is in a deposit position over a rack or stack. When stacking or tiering, only enough backward tilt to stabilize the load shall be used.

p. Operation of the truck

1. If at any time a powered industrial truck is found to be in need of repair, defective, or in any way unsafe, the truck shall be taken out of service until it has been restored to safe operating condition.

2. Fuel tanks shall not be filled while the engine is running. Spillage shall be avoided.

3. No truck shall be operated with a leak in the fuel system; the leak must be corrected.

4. Open flames shall not be used for checking the electrolyte level in storage batteries or the gasoline level in fuel tanks.

q. Maintenance of industrial trucks

1. Any power-operated industrial truck not in safe operating condition shall be removed from service. All repairs shall be made by authorized personnel.

2. No repairs shall be made in Class I, II, and III locations. (The classes indicate the degree of hazard.)

3. Those repairs to the fuel and ignition systems of industrial trucks that involve fire hazards shall be conducted only in locations designated for such repairs.

4. Trucks in need of repairs to the electrical system shall have the battery disconnected prior to such repairs.

5. All parts of any such industrial truck requiring replacement shall be replaced only by parts as safe as those used in the original design.

6. Industrial trucks shall not be altered so that the relative positions of the various parts are different from what they were when originally received from the manufacturer, nor shall they be altered either by the addition of extra parts not provided by the manufacturer or by the elimination of any parts, except as provided in item 12 of this list. Additional counterweighting of fork trucks shall not be done unless approved by the truck manufacturer.

7. Industrial trucks shall be examined before being placed in service, and shall not be placed in service if the examination shows any condition adversely affecting the safety of the vehicle. Such examination shall be made at least daily. Where industrial trucks are used on a round-the-clock basis, they shall be examined after each shift. Defects when found shall be immediately reported and corrected.

8. Water mufflers shall be filled daily or as frequently as necessary to prevent depletion of the supply of water below 75% of the filled capacity. Vehicles with mufflers having screens or other parts that may become clogged shall not be operated while such screens or parts are clogged. Any vehicle that emits hazardous sparks or flames from the exhaust system shall immediately be removed from service and not returned to service until the cause for the emission of such sparks and flames has been eliminated.

9. When the temperature of any part of any truck is found to be in excess of its normal operating temperature, thus creating a hazardous condition, the vehicle shall be removed from service and not returned to service until the cause for such overheating has been eliminated.

10. Industrial trucks shall be kept in a clean condition, free of lint, excess oil, and grease. Noncombustible agents should be used for cleaning trucks. Low-flash-point (below 100 °F) solvents shall not be used. High-flash-point (at or above 100 °F) solvents may be used. Precautions regarding toxicity, ventilation, and fire hazard shall be consonant with the agent or solvent used.

11. Where it is necessary to use antifreeze in the engine cooling system, only those products having a glycol base shall be used.

12. Industrial trucks originally approved for the use of gasoline for fuel can be converted to liquefied petroleum gas fuel provided the complete conversion results in a truck that embodies the features specified for LP or LPS designated trucks. Such conversion equipment shall be approved. The description of the component parts of this conversion system and the recommended method of installation on specific trucks are contained in the "Listed by Report" (39FR23502, June 27, 1974, as amended at 40FR23073, May 28, 1975).

Index

Other Bestsellers of Related Interest

ADVANCED MANUFACTURING TECHNOLOGY—T.H. Allegri, P.E.

T.H. Allegri illustrates many ways that various disciplines have been integrated in order to produce advancements in the technology of manufacturing. Digesting dozens of references, Allegri covers: CIM, CAD, CAE, management information systems, robotics, graphics and simulation, automated materials handling systems, quality control and product life cycle, flexible manufacturing systems, and more. 400 pages, 63 illustrations. Book No. 2746, $44.50 hardcover only

ENGINEERING DESIGN: Reliability, Maintainability, and Testability
—James V. Jones

Today's economy demands low-cost, reliable products. To meet that demand, design engineers must be able to coordinate their efforts with those of end users, test technicians, manufacturing and support personnel, and other engineers. In this comprehensive guide, management and logistics expert Jim Jones presents a total, field-tested plan that will help you keep your customers satisfied and increase your chances for success. 334 pages, 188 illustrations. Book No. 3151, $42.95 hardcover only

DESIGN GUIDELINES FOR SURFACE MOUNT TECHNOLOGY—Vern Solberg

Increased production volumes . . . more efficient, cost effective products . . . improved manufacturing efficiency . . . reduced size of electronic products—these are just a few of the advantages surface mount technology (SMT) can bring to your manufacturing process. This new guide offers the most detailed coverage available on the design, manufacture, and testing of substrate assemblies using SMT. 192 pages, 166 illustrations. Book No. 3199, $52.00, hardcover only

SOLDER PASTE TECHNOLOGY: Principles and Applications
—Colin C. Johnson and Joseph Kevra, Ph.D.

This reference offers you comprehensive discussion of theory and application of solder pastes with particular emphasis on how they affect the electronics industry. From metallurgy, rheology, and fine particle measurement to chemistry, screening, and dispensing technologies, it's all covered here. If you are involved in any aspect of solder paste use, you will find this book indispensable. 288 pages, 162 illustrations. 4 pages color. Book No. 3203, $46.95 hardcover only

THE AUTOMATED FACTORY HANDBOOK: Technology and Management
—Edited by David I. Cleland and Bopaya Bidanda

This outstanding collection written by experts in the field shows you how to successfully design and use computer-integrated manufacturing systems. All the tools, advanced management technology, and personnel management concepts needed to run an automated factory are discussed. Case studies illustrate important concepts. 832 pages, 359 illustrations. Book No. 3296, $69.95 hardcover only

HUMAN FACTORS IN INDUSTRIAL DESIGN: The Designer's Companion
—John H. Burgess

This book presents a nonscientific introduction to human-factor considerations. It's directed specifically to the everyday needs of industrial designers, emphasizing the importance of creating products and equipment that are both safe and useful to their human users. Burgess explains the methods employed by human-factors specialists in determining the significance of human dimensions, capabilities, and limitations in designing everything from small hand tools to large, complex systems. 218 pages, illustrated. Book No. 3356, $27.95 hardcover only

Prices Subject to Change Without Notice.

Look for These and Other TAB Books at Your Local Bookstore

To Order Call Toll Free 1-800-822-8158
(in PA, AK, and Canada call 717-794-2191)

or write to TAB Books, Blue Ridge Summit, PA 17294-0840.

Title	Product No.	Quantity	Price

☐ Check or money order made payable to TAB Books

Charge my ☐ VISA ☐ MasterCard ☐ American Express

Acct. No. _____ Exp. _____

Signature: _____

Name: _____

Address: _____

City: _____

State: _____ Zip: _____

Subtotal $ _____

Postage and Handling
($3.00 in U.S., $5.00 outside U.S.) $ _____

Add applicable state and local
sales tax $ _____

TOTAL $ _____

TAB Books catalog free with purchase; otherwise send $1.00 in check or money order and receive $1.00 credit on your next purchase.

Orders outside U.S. must pay with international money order in U.S. dollars.

TAB Guarantee: If for any reason you are not satisfied with the book(s) you order, simply return it (them) within 15 days and receive a full refund. **BC**

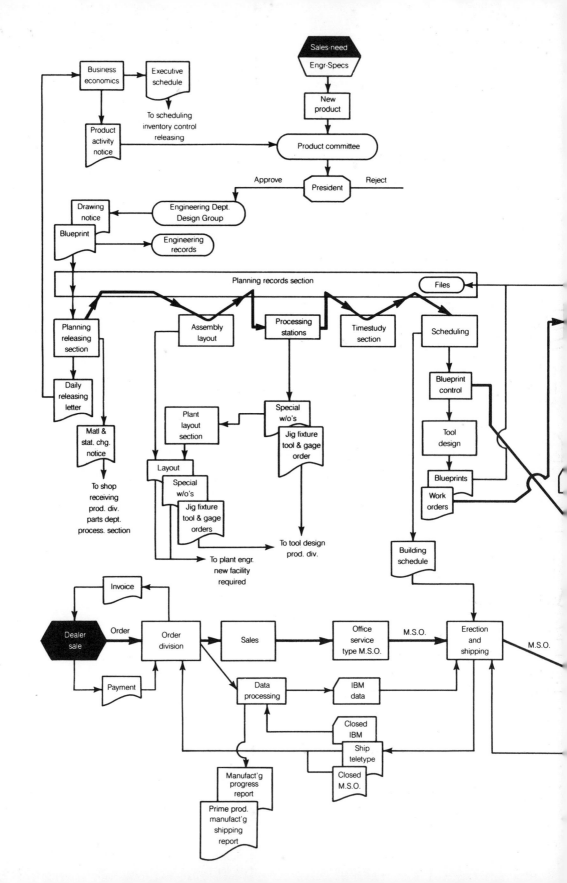